NASA SP–4402

ORIGINS OF NASA NAMES

Helen T. Wells, Susan H. Whiteley,
and Carrie E. Karegeannes

The NASA History Series

Scientific and Technical Information Office   1976
NATIONAL AERONAUTICS AND SPACE ADMINISTRATION
Washington, D.C.

Library of Congress Cataloging in Publication Data

Wells, Helen T
    Origins of NASA names.

    (NASA SP ; 4402)
    Includes bibliographical references and index.
    Supt. of Docs. no.: NAS 1.21:4402
    1. United States. National Aeronautics and Space Administration. 2. Astronautics—United States.
3. Aeronautics—United States. I. Whiteley, Susan H., joint author. II. Karegeannes, Carrie E., joint
author. III. Title. IV. Title: NASA names. V. Series: United States. National Aeronautics and Space
Administration. NASA SP ; 4402.
TL521.312.W45   629.4'0973   76–608131

FOREWORD

This book was started many years ago. From time to time, the work was interrupted in favor of tasks that seemed more pressing. Meanwhile, the number of names generated by NASA continued to grow, and the work to be done increased. Now it has been brought to completion, and I am happy to offer it to the public. From the number of times the staff has consulted the manuscript to answer telephone queries, the publication should prove useful.

MONTE D. WRIGHT
Director, NASA History Office
June 1975

CONTENTS

Page

Foreword   iii

Preface   vii

Note   ix

I. Launch Vehicles   1

II. Satellites   29

III. Space Probes   81

IV. Manned Space Flight   95

V. Sounding Rockets   119

VI. NASA Installations   135

Appendix A. Selected List of Abbreviations, Acronyms, and Terms.   161

Appendix B. International Designation of Spacecraft.   167

Appendix C. NASA Major Launch Record, 1958–1974   169

Appendix D. NASA Naming Committees.   183

Reference Notes   185

Index.   211

The Authors   229

PREFACE

This book was designed to answer questions about the origins of NASA-associated names. The impetus for its preparation came from the Johnson Space Center Historian, James M. Grimwood, who called attention to the need for such a compilation. If, besides answering specific questions, the book raises further questions or stimulates the reader to delve further into the subject of space exploration, it will have served its purpose well.

Names given to spaceflight projects and programs have originated from no single source or method. Some have their foundations in mythology and astrology, some in legend and folklore. Some have historic connotations. Some are based on a straightforward description of their mission, often resulting in acronyms. (As Webster puts it, an acronym is a "word formed from the initial letters or syllables of the successive parts of a compound term"; hence, "TIROS" for "Television and Infra-Red Observation Satellite.") Some grew out of a formal process within NASA under the NASA Project Designation Committee. Others evolved somewhat casually and were officially adopted after their use had become widespread. Many others, of course, were originated by non-NASA sources when ongoing projects were transferred to NASA from other agencies.

Parts I through V list names of launch vehicles, spacecraft, manned spaceflight programs, and sounding rockets. Some of these were the primary responsibility of NASA. Some names apply to projects for which NASA shared responsibility or had a major support role—for instance, the international Alouette satellites. Some names apply to hardware that NASA purchased from another agency, such as the Air Force Agena launch vehicle stage. Part VI lists NASA field installations and gives the origins of their names.

This study is limited to names of approved projects through 1974; it does not include names of numerous projects which have been or are being studied or projects that were canceled or postponed before reaching actual flight—such as the Nova large launch vehicle. It does not attempt to record the history of the listed projects except as it may be related to the naming process, nor does it attempt to describe the projects and hardware beyond a nontechnical statement of mission or function. It does, however, present the origins of each name, answering as far as possible who or what organization devised the name, when they adopted it, and what the reasoning was for its selection. The information about each name is as specific as the available documentation could provide. Because of the passage of time, the multiplicity of organizations participating, and the unavailability of full written documentation, it is, we regret, inevitable that some persons deserving of credit in the naming processes have been overlooked.

So many persons have provided information that it would be impossible to acknowledge each one's contribution. Reference notes attempt to credit specific assistance on particular points. Special mention is due the historians, their staffs, and historical monitors at NASA Centers, who coordinated local research, and to Dr. Eugene M. Emme, NASA Historian, and Dr. Frank W. Anderson, Jr., Publications Manager, who gave invaluable guidance in organizing and editing the manuscript. Arthur G. Renstrom of the Library of Congress was very helpful in finding illustrative materials, and NASA Archivist Lee D. Saegesser spent many hours tracking down historical photographs of spacecraft and vehicles. Sources of the photographs of mythological figures are listed at the end of the Reference Notes.

Comments and additional information on the origins of NASA names will always be welcomed.

HTW
SHW
CEK

NOTE

For consistency and to avoid confusion, the numerical designations of spacecraft within the text conform to the arabic numeral system. Until 1969, NASA chose roman numerals to designate successful flight missions, although there were notable exceptions. Italics indicate spacecraft that have attained orbit, space probes that have achieved an altitude above 64 000 kilometers, and all manned suborbital flights. Spacecraft launch failures retain their preflight letter designations.

No single system of numbering spacecraft and launch vehicles has been followed by NASA through the years, and often two or more designations have existed for one spacecraft. Usually, however, spacecraft in a series are given letter designations in alphabetical order before launch and successful launches within a series are numbered consecutively with arabic numerals.

Many satellites and space probes have followed this pattern and most launch vehicles have been numbered separately. There were exceptions: launches that failed sometimes upset the numbering sequence (for example, *Pioneer 3* followed *Pioneer 1*) and designations for spacecraft within a series at times did not appear to follow a given sequence (ATS–B was launched before ATS–A; *OAO 2* was known as OAO–A2 before launch). Launch vehicle development flights were numbered consecutively, including suborbital flights (Atlas-Centaur 2; Atlas-Centaur 3, suborbital; Atlas-Centaur 4).

The numbering systems for the first three manned spaceflight programs—as the new space agency developed approaches in an evolving field—were not consistent. In addition to the overall flight number, each manned flight had a separate designation, named and numbered in sequence, for the launch vehicle combination that was employed. Except for flights in the Mercury program and unmanned flights in the Gemini program, these secondary designations were not official names, but were used by NASA for reference (in the Gemini program) or launch vehicle designation (Apollo). For example, *Freedom 7* was also known as *Mercury-Redstone 3* and *Friendship 7* as *Mercury-Atlas 6*.

In the Mercury program, the choice of the number "7" by the original seven Mercury astronauts precluded the use of roman numerals for the spacecraft. Project Gemini was the only manned program to use roman numerals, and even its early unmanned flights were named by the Mercury system (*Gemini-Titan 1, Gemini-Titan 2*). In the Apollo program, each mission was assigned an overall number and each command and service module and lunar module was given a separate number designating a specific flight unit (for example, "CSM–108" and "LM–6" designated the specific *Apollo 12* modules). Each Apollo launch vehicle was assigned a flight number that indicated both the vehicle model and the specific vehicle used on that mission, such as "AS–201" and "AS–507." The "200" and "500" series referred to Saturn IB and Saturn V launch vehicles; "AS" to "Apollo Saturn." The 10 Saturn I development flights, on the other hand, were designated "Saturn Apollo."

In addition, each spacecraft or piece that separately entered earth orbit was given a number and letter designation, according to the international designation system (see Appendix B). Spacecraft that separated while in orbit or after leaving earth orbit also were given designations (for example, the *Apollo 15 Subsatellite*, ejected into lunar orbit from the Apollo command module, and the Soviet softlanding capsules that descended to the Martian surface from the *Mars 2* and *3* space probes).

I

LAUNCH VEHICLES

Dr. Robert H. Goddard with the world's first liquid-propellant rocket, launched 16 March 1926 at Auburn, Massachusetts.

LAUNCH VEHICLES

Launch vehicles are the rocket-powered systems that provide transportation from the earth's surface into the environment of space. In the early days of the U.S. civilian space program the term "launch vehicle" was used by NASA in preference to the term "booster" because "booster" had been associated with the development of the military missiles. "Booster" now has crept back into the vernacular of the Space Age and is used interchangeably with "launch vehicle."

In 1971 NASA managed five launch vehicles in the National Launch Vehicle Program: Scout, Thor-Delta, Atlas-Centaur, Saturn IB, and Saturn V. In 1974 a new combination, the Titan-Centaur, launched its first satellite. Performance capability of these vehicles varied greatly, ranging from Scout, which was used to launch small scientific payloads, to Saturn V, which launched manned Apollo missions into circumlunar flight. Beginning in the 1980s, NASA's reusable Space Shuttle was scheduled to replace many expendable boosters for orbiting satellites and manned missions (see Space Shuttle in Part IV).

Names listed in this section include designations of launch vehicles and major vehicle stages, or sections, that are used or have been used in the past by NASA. Nearly all the names came from the military services or the Department of Defense, which traditionally turned to ancient mythology in selecting names for ballistic missiles and space boosters.

Thor-Able

ABLE. The Able upper stage was one of several derived in 1958 by the Department of Defense's Advanced Research Projects Agency, Douglas Aircraft Company, and Space Technology Laboratories from Vanguard launch vehicle components. It was used with Thor or Atlas first stages. The name signified "A" or "first" (from military phonetic communications practice of stipulating key words beginning with each letter of the alphabet).[1] (See Delta.)

Thor-Agena B before launch from the Pacific Missile Range, at left. Opposite, the Gemini Agena Target Vehicle as seen from the Gemini 8 spacecraft during approach for rendezvous and docking 16 March 1966.

AGENA. An upper-stage launch vehicle used in combination with Thor or Atlas first stages, Agena originally was developed for the U.S. Air Force by Lockheed Missiles Systems Division (now Lockheed Missiles & Space Company). The Department of Defense's Advanced Research Projects Agency (ARPA) proposed to name the stage in 1958 for the star Agena in the constellation Centaurus because the rocket was an upper stage "igniting in the sky."[1] "Agena" first appeared in the *Geography of the Heavens*, published in the 1800s by the "popularizing Connecticut astronomer" Elija H. Burritt, and was preserved in American dictionaries as the popular name for the

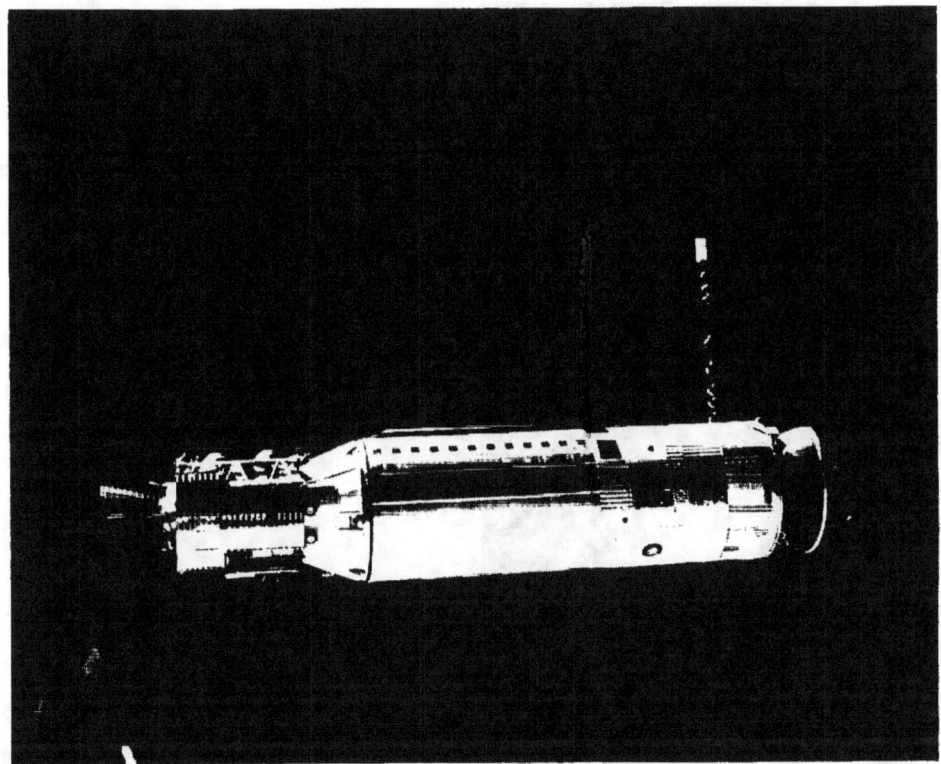

star Beta Centauri.[2] Burritt was thought to have coined the name from *alpha* and *gena* ("the knee") because he had located the star near the "right foreleg" of the constellation.[3] Lockheed approved the choice of the name since it followed Lockheed's tradition of naming aircraft and missiles after stellar phenomena—such as the Constellation aircraft and Polaris intercontinental ballistic missile.[4] ARPA formally approved the name in June 1959.[5]

Agena A, the first version of the stage, was followed by the Agena B, which had a larger fuel capacity and engines that could restart in space. The later Agena D was standardized to provide a launch vehicle for a variety of military and NASA payloads.[6] NASA used Atlas-Agena vehicles to launch large earth satellites as well as lunar and interplanetary space probes; Thor-Agena vehicles launched scientific satellites, such as OGO (Orbiting Geophysical Observatory) and Alouette, and applications satellites, such as the *Echo 2* communications satellite and Nimbus meteorological satellites. In Project Gemini the Agena D, modified to suit the specialized requirements of space rendezvous and docking maneuvers, became the Gemini Agena Target Vehicle (GATV).

Launch of Mercury-Atlas carrying Astronaut John H. Glenn, Jr., in Friendship 7 for the first U.S. manned orbital space flight 20 February 1962.

Atlas (courtesy Library of Congress)

Models of Atlas upper-stage configurations. Left to right: Atlas, Atlas-Agena, Mercury-Atlas, Atlas-Centaur, Atlas-Able.

ATLAS. The Atlas launch vehicle was an adaptation of the U.S. Air Force Atlas intercontinental ballistic missile. The modified Atlas launched the four manned orbital flights in Project Mercury and NASA used it with the Agena or Centaur upper stages for a variety of unmanned space missions.

Early in 1951 Karel J. Bossart, head of the design team at Convair (Consolidated Vultee Aircraft Corporation) that was working on the missile project for the Air Force, decided the project (officially listed as MX–1593) should have a popular name. He asked some of his staff for ideas and they considered several possibilities before agreeing upon "Atlas"—Bossart's own suggestion. The missile they were designing would be the biggest and most powerful yet devised. Bossart recalled that Atlas was the mighty god of ancient Greek mythology who supported the world on his powerful shoulders. The appropriateness of the name seemed confirmed by the fact that the parent company of Convair was the Atlas Corporation.[1] The suggestion was submitted to the Air Force and was approved by the Department of Defense Research and Development Board's Committee on Guided Missiles in August 1951.[2]

The Atlas-Centaur, a high-energy vehicle for launching medium-weight spacecraft into planetary or synchronous orbits, could put 4700 kilograms into 555-kilometer orbit or 1810 kilograms into transfer orbit for a synchronous orbit.[3] (See Centaur.)

9

BIG JOE. ''Big Joe'' was the name of a single Atlas booster and its test flight. Part of Project Mercury, Big Joe tested a full-scale Mercury capsule at full operational speed for the critical reentry into the earth's atmosphere. It was a key test of the heatshield, in preparation for Mercury's manned orbital space flights. The name, which developed in 1958, was attributed to Maxime A. Faget, then at Langley Research Center. It was a logical progression from the previously named Little Joe, a smaller test booster for demonstration flight tests in Project Mercury.[1] (See Little Joe.)

Big Joe before launch at Cape Canaveral.

Atlas-Centaur

CENTAUR. Centaur was known from 1956 to 1958 simply as the "high-energy upper stage" because it proposed to make first use of the theoretically powerful but problem-making liquid hydrogen as fuel. The stage was named in November 1958 when the Department of Defense's Advanced Research Projects Agency (ARPA) awarded the initial contract for six research and development flight-test vehicles to Convair/Astronautics Division of General Dynamics Corporation. The Centaur stage was required to increase the payload capability of the Atlas and to provide a versatile second stage for use in complex space missions. Krafft Ehricke of General Dynamics, who conceived the vehicle and directed its development, proposed the name and ARPA approved it. The name derived from the legendary Centaur, half man and half horse. The horse portion represented the "workhorse" Atlas, the "brawn" of the launch vehicle; the man represented the Centaur—

11

which, containing the payload and guidance, was in effect the "brain" of the Atlas-Centaur combination. Eugene C. Keefer of Convair was credited with proposing the name to Ehricke.[1]

NASA, which received management responsibility for the Atlas-Centaur, used the launch vehicle in the Intelsat IV series of comsats and the Surveyor series of space probes. Centaur was also used to launch some of the larger satellites and space probes—such as *OAO 2* and *3, ATS 5,* and the heavier Mariner and Pioneer space probes—and was mated with the Air Force Titan III for the heavier payloads flown in the mid-1970s. NASA launched the U.S.-German *Helios I* into orbit of the sun on a Titan IIIE–Centaur on 10 December 1974. (See also Atlas and Titan.)

Centaur figures (courtesy of the Library of Congress)

DELTA. When NASA was formed in 1958 it inherited from the Department of Defense's Advanced Research Projects Agency (ARPA) the booster programs using combinations of Thor or Atlas boosters with Vanguard upper stages. The first of these upper-stage configurations was designated "Able." [1] The Delta was similar to the previous Thor-based combinations and was a fourth—or "D"—version. Milton W. Rosen of NASA was responsible for the name. He had been referring to the combination as "Delta," which became the firm choice in January 1959 when a name was required because NASA was signing a contract for the booster. The vehicle was variously called "Delta" and "Thor-Delta." [2]

Over the years the Thor-Delta was repeatedly uprated by additions and modifications. The liftoff thrust of the Thor first stage was increased in 1964 by adding three strapped-on solid-propellant rocket motors. With the Delta second stage, the launch vehicle was called "thrust-augmented Delta" (TAD). In 1964 NASA undertook upgrading the Delta capability by enlarging the second-stage fuel tanks. When this more powerful version—

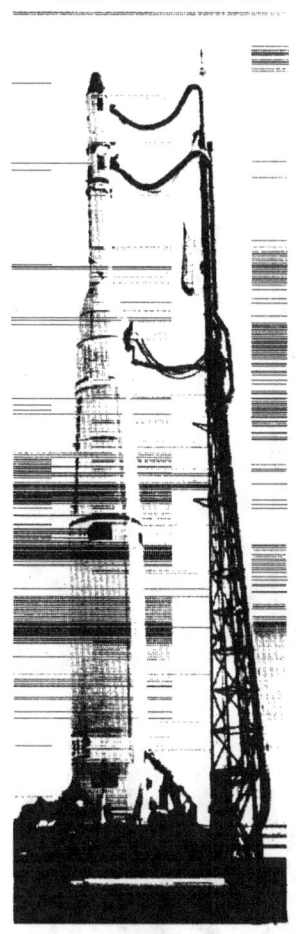

Thor-Delta

Thrust-augmented im-
proved Delta

Straight Eight Thor-Delta
with nine strap-on rockets

introduced in 1965 and designated "improved Delta"—was used with the thrust-augmented Thor first stage, the vehicle was called "thrust-augmented improved Delta" (TAID). In 1968 NASA incorporated an elongated Thor first stage with added fuel capacity for heavier payloads, and the three strapped-on motors were uprated. This version, with the improved Delta second stage, was called "long-tank thrust-augmented Thor-Delta (LTTAT-Delta), or "thrust-augmented long-tank Delta." [3]

The "Super Six" version, with six strap-on Castor rockets for extra thrust, was first used in 1970, and nine strap-ons went into use in 1972. A more powerful third stage, TE–364–4, was also introduced in 1972, as was the "Straight Eight" Thor-Delta, with 2.4-meter (8-foot) diameter for all three stages including the fairing. The wider fairing could accommodate larger spacecraft. [4]

In 1960 the Thor-Delta placed 60 kilograms in a 1600-kilometer orbit. By

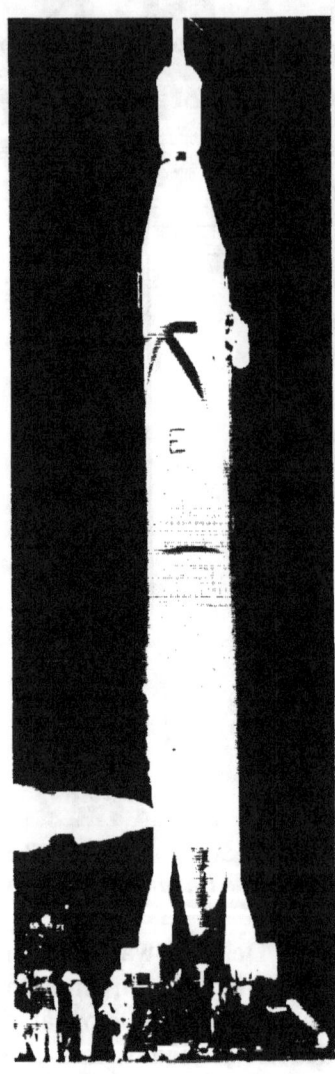

Juno (courtesy of the Library of Congress)

Juno I, above, on the launch pad at Cape Canaveral before launching Explorer 1 *on 31 January 1958. At right, Little Joe launch in a test of the Mercury spacecraft.*

the end of 1974, the vehicle could launch a 700-kilogram spacecraft into orbit for transfer to a 35 500-kilometer synchronous orbit, an 1800-kilogram payload into a 185-kilometer orbit, or 386 kilograms on a trajectory to Mars or Venus.[5]

The economical, reliable Thor-Delta was a workhorse vehicle used for a wide range of medium satellites and small space probes in two-stage or three-stage combinations, with three, six, or nine strap-on thrust-augmentor rockets. Among its many credits were meteorological satellites (Tiros, TOS), communications satellites (Echo, Telstar, Relay, Syncom, Intelsat), scientific satellites (Ariel, Exploror, OSO), and the Earth Resources Satellite *ERTS 1*. The vehicle's first three-satellite launch orbited *NOAA 4, OSCAR 7,* and *INTASAT* on 15 November 1974.

JUNO. Juno I and Juno II were early launch vehicles adapted from existing U.S. Army'missiles by the Army Ballistic Missile Agency (ABMA) and the Jet Propulsion Laboratory (JPL). The ancient Roman goddess Juno, queen of the gods, was the sister and wife of Jupiter, king of the gods. Since the new launch vehicle was the satellite-launching version of the Jupiter C (Jupiter Composite Reentry Test Vehicle), the name Juno was suggested by Dr. William H. Pickering, JPL Director, in November 1957. Army officials approved the proposal and the name was adopted.[1]

Juno I, a four-stage configuration of the Jupiter C, orbited the first U.S. satellite, *Explorer 1,* 31 January 1958. The "UE" painted on the Redstone first stage of that Juno I indicated that the Redstone was No. 29 in a series of launches. The ABMA code for numbering Redstone boosters was based on the word "HUNTSVILLE," with each letter representing a number, after deletion of the second "l" to avoid confusion:

H U N T S V I L E [2]
1 2 3 4 5 6 7 8 9.

Later that year, at the request of the Department of Defense's Advanced Research Projects Agency, ABMA and JPL designed the Juno II, which was based on the Jupiter intercontinental ballistic missile and had the upper stages of the Juno I. Responsibility for Juno II was transferred to NASA after its establishment 1 October 1958. Juno II vehicles launched three Explorer satellites and two Pioneer space probes. "Juno V" was the early designation of the launch vehicle that became the Saturn I.[3]

LITTLE JOE. A relatively simple and inexpensive launch vehicle, Little Joe was designed specifically to test the Mercury spacecraft abort system in a series of suborbital flights. Based on a cluster of four solid-propellant rocket motors, as conceived by Langley Research Center's Maxime A. Faget and

Paul E. Purser, the booster acquired its name in 1958 as Faget's nickname for the project gradually was adopted. The configuration used in the tests added four Recruit rockets, but the original concept was for four Pollux rocket motors fired two at a time—a pair of twos. "Since their first cross-section drawings showed four holes up, they called the project 'Little Joe,' from the crap game throw of a double deuce on the dice. . . . The appearance on engineering drawings of the four large stabilizing fins protruding from its airframe also helped to perpetuate the name Little Joe had acquired." [1] Little Joe II was similar in design and was used to check out the Apollo spacecraft abort system.

REDSTONE. Predecessor of the Jupiter and Juno rockets, Redstone was a battlefield missile developed by the U.S. Army and adapted for use by NASA as a launch vehicle for suborbital space flights in Project Mercury. After being called various nicknames, including "Ursa" and "Major," the missile was officially named "Redstone" 8 April 1952 for the Army installation Redstone Arsenal at Huntsville, Alabama, where it was developed.[1] The name of the Arsenal, in turn, referred to the rock and soil at Huntsville.[2]

On 5 May 1961, the Redstone launched the first U.S. astronaut, Alan B. Shepard, Jr., into suborbital flight on the *Freedom 7* in Project Mercury.

Launch of Mercury-Redstone from Cape Canaveral 5 May 1961, carrying Astronaut Alan B. Shepard, Jr., on the first U.S. manned space flight.

Saturn I being readied for launch from Cape Kennedy with boilerplate model of the Apollo spacecraft.

SATURN I, SATURN IB. Evolution of nomenclature for the Saturn family of launch vehicles was one of the most complex of all NASA-associated names. On 15 August 1958 the Department of Defense's Advanced Research Projects Agency (ARPA) approved initial work on a multistage launch vehicle with clustered engines in a 6.7-million-newton-thrust (1.5-million-pound-thrust) first stage. Conceived by designers at the Army Ballistic Missile Agency (ABMA), the vehicle was unofficially known as "Juno V." (Juno III and Juno IV were concepts for space vehicles to follow Juno II but were not built.) [1]

In October 1958 Dr. Wernher von Braun, the Director of ABMA's Development Operations Division, proposed the Juno V be renamed "Saturn," and on 3 February 1959 ARPA officially approved the name change. The name "Saturn" was significant for three reasons: the planet Saturn appeared brighter than a first-magnitude star, so the association of this name with such a powerful new booster seemed appropriate; Saturn was the next planet after Jupiter, so the progression was analogous to ABMA's progression from missile and space systems called "Jupiter"; and Saturn was the name of an ancient Roman god, so the name was in keeping with the U.S. military's custom of naming missiles after mythological gods and heroes. [2]

Throughout the second half of 1959, studies were made of possible upper stages for the new Saturn vehicle. The interagency Saturn Vehicle Evaluation Committee,* considered many combinations, narrowing the choice to

---

*Chaired by NASA's Abe Silverstein and often referred to as "the Silverstein Committee," the committee was composed of representatives of NASA, ARPA, DOD, and USAF.

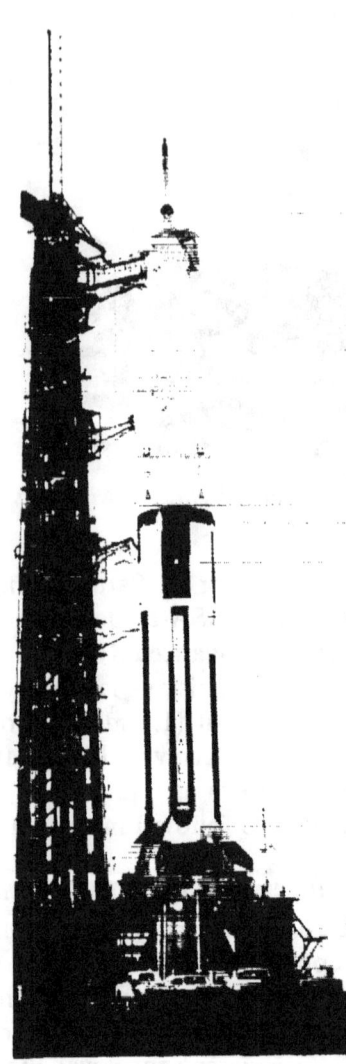

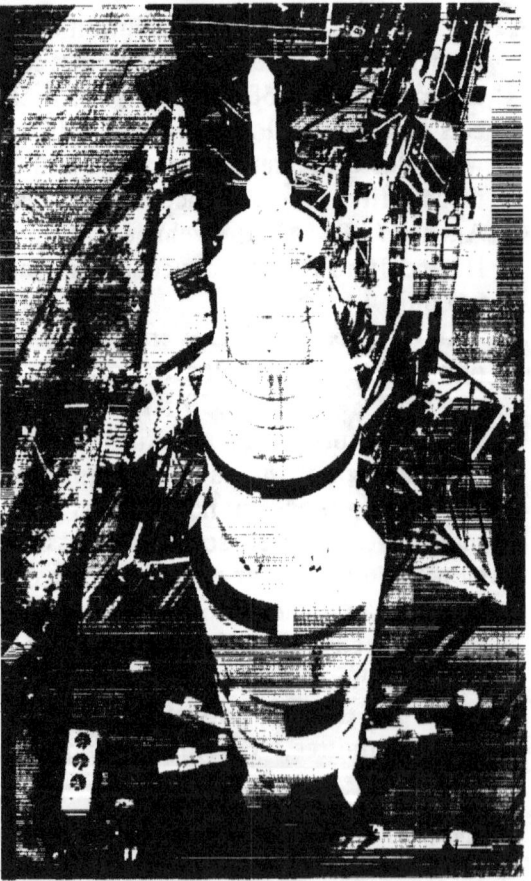

Saturn IB, above, on the launch pad at Complex 34 and Saturn of mythology (courtesy of the Library of Congress). At right, the rollout of Saturn V for the launch of Apollo 8 to the moon.

design concepts labeled "Saturn A," "Saturn B," and "Saturn C." In December 1959, following the recommendation of this committee, NASA authorized building 10 research and development models of the first "C" version, or "Saturn C–1" design proposal.[3] For the time being the booster was called "Saturn C–1."

In the meantime Saturn became a NASA project and also had become an important link with the Nation's manned lunar program, Project Apollo. In 1962, NASA decided a more powerful version of the Saturn C–1 would be needed to launch Apollo lunar spacecraft into earth orbit, to prepare and train for manned flights to the moon later in the 1960s. NASA called this launch vehicle "Saturn C–1B."[4] In February 1963, NASA renamed these vehicles. At the suggestion of the NASA Project Designation Committee, Saturn C–1 became simply "Saturn I" and the Saturn C–1B, "Saturn IB."[5] The Saturn IB was composed of the S–IB first stage, a modified version of the S–I first stage that could develop 7.1 million newtons (1.6 million pounds) of thrust by 1973, and the S–IVB second stage, an uprated version of the S–IV stage that could develop 1 million newtons (230 000 pounds) of thrust.

On 9 June 1966 NASA changed the name of the Saturn IB to "Uprated Saturn I." The redesignation was suggested to the Project Designation Committee by Dr. George E. Mueller, NASA Associate Administrator for Manned Space Flight. "The Committee agreed with Dr. Mueller that the booster is acutally an uprated Saturn I and should be so called."[6] In December 1967, however, NASA decided to return to the use of the simpler term, "Saturn IB." The proposal was made by the Office of Manned Space Flight and approved by Administrator James E. Webb.[7]

The Saturn IB launched the first manned Apollo spacecraft, *Apollo 7,* on successful flight 11 October 1968 and, after the completion of the Apollo program, launched three missions to man the Skylab Orbital Workshop in 1973. It was scheduled to launch the American crew in the July 1975 U.S.– U.S.S.R. Apollo-Soyuz Test Project docking mission.

SATURN V. In January 1962 NASA initiated development of the large launch vehicle for Project Apollo manned lunar flight. The vehicle selected was the Saturn C–5, chosen after six months of studying the relative merits of Saturn C–3, C–4, and C–5 designs.[1] These designs were all based on a large clustered-engine first stage but with various combinations of upper stages. The numerical designation followed the sequence established with the Saturn C–1 (see Saturn I, where the origin of the name "Saturn" also is explained).

Alternately referred to in 1962 as "Advanced Saturn," the Saturn C–5 was renamed early the following year. Nominations were submitted to the NASA Project Designation Committee as well as proposed by the Committee members themselves. After considering many alternate names—the leading contender for a while was "Kronos"—the Committee suggested, through Assistant Administrator for Public Affairs George L. Simpson, Jr., to NASA Associate Administrator, Dr. Robert C. Seamans, Jr., that the new name be "Saturn V." [2] The recommendation was approved and the new name adopted early in February 1963. [3]

The final configuration of the Saturn V comprised the S–IC first stage with 34-million-newton (7.7-million-pound) thrust, the S–II second stage with 5.1-million-newton (1.2-million-pound) thrust, and the S–IVB stage of the Saturn IB. [4]

On 21 December 1968 the Saturn V launched *Apollo 8*, the first manned Apollo spacecraft to escape the earth's gravitational field, into flight around the moon. Saturn V launches through *Apollo 17* in December 1972 put 27 men into lunar orbit, 12 of them landing on the moon to explore its surface. On 14 May 1973 the Saturn V orbited the first U.S. experimental space station, the *Skylab 1* Orbital Workshop, which was manned by three successive three-man crews during the year.

SCOUT. The Scout launch vehicle was named in mid-1958 by William E. Stoney, Jr., prominent in development of the vehicle at NACA Langley Aeronautical Laboratory (later NASA Langley Research Center). He thought of the name as a parallel to "Explorer," a name being given to a series of spacecraft. "Scout" seemed appropriate for a vehicle with payloads performing similar tasks—"scouting the frontiers of space environment and paving the way" for future space exploration. [1]

Smallest of the basic launch vehicles, Scout was designed at Langley as a reliable, relatively inexpensive launch vehicle for high-altitude probes, reentry experiments, and small-satellite missions. Among the satellites it launched were scientific satellites such as Explorers and international satellites such as the San Marco series. It was the only U.S. satellite launch vehicle to use solid propellants exclusively; the stages for Scout had grown out of the technology developed in the Polaris and Minuteman programs. The Air Force, which used Scout to launch Department of Defense spacecraft, called its version "Blue Scout."

Scout usually consisted of four stages and could put 186 kilograms into a 555-kilometer orbit. [2] The first stage, "Algol," was named for a star in the constellation Perseus; the second stage, "Castor," for the "tamer of the horses" in the constellation Gemini; the third stage, "Antares," for the

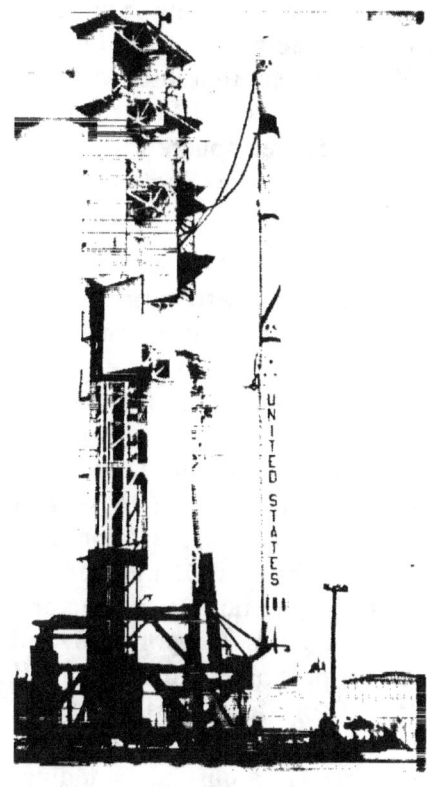

Scout on the launch pad at Wallops Flight Center, at left. Below, Shotput ready at Wallops.

brightest star in the constellation Scorpio; and the fourth stage, "Altair," for a star in the constellation Aquila.[3] In June 1974 a new Scout E, incorporating a solid-fueled rocket motor in a fifth stage and adaptable for highly eccentric orbits, launched the *Hawkeye 1* Explorer satellite.[4]

SHOTPUT. A special-purpose composite rocket to test balloon-satellite ejection and inflation in space, Shotput was used in five launches from Wallops Station in 1959 and 1960 in tests of the *Echo 1* satellite payload. It was also used to test the Italian San Marco satellite in suborbital flights. The solid-propellant Shotput vehicle consisted of a first-stage Sergeant rocket boosted by two Recruit rockets and a second stage X–248 rocket that later was used as the third stage of the Delta launch vehicle.[1] Shotput launched the balloon payload to a 400-kilometer altitude, where the packaged sphere was ejected from the vehicle's nose and inflated above the atmosphere. Shotput was so named because it "tossed" the Echo sphere up above the earth's atmosphere in a vertical trajectory.[2]

THOR. Adapted for use as a launch vehicle in combination with various upper stages, Thor was originally developed as a U.S. Air Force intermediate-range ballistic missile by Douglas Aircraft Company. The name, which came into use in 1955,[1] derived from the ancient Norse god of thunder—"the strongest of gods and men."[2]

The origin of the name has been traced back to Joe Rowland, Director of Public Relations at the Martin Company, who was assigned to suggest names for Martin's new intercontinental ballistic missile in preparation for a meeting at Air Research and Development Command (ARDC) Headquarters. At the meeting were to be representatives of other missile contractors, Convair/Astronautics Division of General Dynamics Corporation and Douglas Aircraft Company. Of Rowland's list of proposed names, "Titan" was the one preferred by his colleagues, with "Thor" as second choice. At the ARDC meeting, the first-choice "Titan" was accepted as the appropriate name for the Martin Company's project. Through a misunderstanding, Douglas had prepared no name to propose for its missile. Rowland—with "Titan" now firm for his company's project—offered his alternate "Thor" to Donald Douglas, Jr. Douglas and his Vice President of Public Relations agreed it was an attractive name and proposed it to ARDC officials; it was officially adopted.[3]

NASA used Thor as a first stage with both Agena and Delta upper stages. The Air Force-developed "thrust-augmented Thor" (TAT), with three added solid-propellant rocket motors strapped on the base of the Thor, also was used with both Agena and Delta upper stages. When TAT was used with Agena, the configuration was called "thrust-augmented Thor-Agena";[4] with Delta, the vehicle was known as "thrust-augmented Delta" (TAD) or "thrust-augmented Thor-Delta" (TAT–Delta).

In 1966 the Air Force procured a new version of the Thor first stage, elongated to increase fuel capacity, for heavier payloads—the "long-tank thrust-augmented Thor" (LTTAT), sometimes also called "Thorad." LTTAT used with an Agena upper stage was called "long-tank thrust-augmented Thor-Agena" or "Thorad-Agena."[5] With Delta, it was "long-

Thor-Delta, above, in countdown for the Telstar 1 *launch. At right, the long-tank Thor-Delta poised for launch from the ETR. The Norse god Thor at left (courtesy of the Library of Congress).*

tank thrust-augmented Thor-Delta.'' NASA began using the long-tank Thor with the improved Delta second stage in 1968, going to six strap-on rockets for extra thrust in 1970 and introducing nine strap-on rockets in 1972. Combinations varied according to the performance needed for the mission. (See Delta.)

LAUNCH VEHICLES

TITAN. The Titan II launch vehicle was adapted from the U.S. Air Force intercontinental ballistic missile to serve as the Gemini launch vehicle in NASA's second manned spaceflight program. Originating in 1955, the name "Titan" was proposed by Joe Rowland, Director of Public Relations at the Martin Company, producer of the missile for the Air Force. Rowland was assigned the task of suggesting possible names for the project, requested of Martin by the Air Research and Development Command. Of the list of possible names, "Titan" was preferred.[1] He took the name from Roman mythology: the Titans were a race of giants who inhabited the earth before men were created. ARDC approved the nomination and "Titan" became the official name. When the improved version of the missile was developed, the original Titan came to be known as Titan I and the second, Titan II.[2] Titan II was chosen as the Gemini launch vehicle because greater thrust was required to orbit the three-and-a-half-metric-ton Gemini spacecraft; also its storable fuels promised the split-second launch needed for rendezvous with the target vehicle.

The Titan III—an improved Titan II with two solid-propellant strap-on rockets—was developed for use by the Air Force as a standardized launch vehicle that could lift large payloads into earth orbit. NASA contracted for Titan III vehicles for a limited number of missions to begin in the mid-1970s: ATS satellites would require the Titan IIIC vehicles and HEAO satellites, the Titan IIID configuration. Interplanetary missions requiring high-velocity escape trajectories—the Viking Mars probes and Helios solar probes—began using the Titan III–Centaur configuration on completion of the Centaur integration program in 1974. A Titan IIIE–Centaur launched *Helios 1* into orbit of the sun 10 December 1974.

In 1974 the Titan IIIC—which launched *ATS 6* on 30 May 1974—could put an 11 820-kilogram payload into a 555-kilometer orbit or 1500 kilograms into synchronous orbit. The Titan IIIE–Centaur could launch 5135 kilograms into an earth-escape orbit or 3960 kilograms to Mars or Venus.[3]

Gemini-Titan 1 on the launch pad at Cape Kennedy during prelaunch tests (far left) and Titan IIIE–Centaur. At right, Ocean, one of the Titans of mythology (courtesy of the Library of Congress).

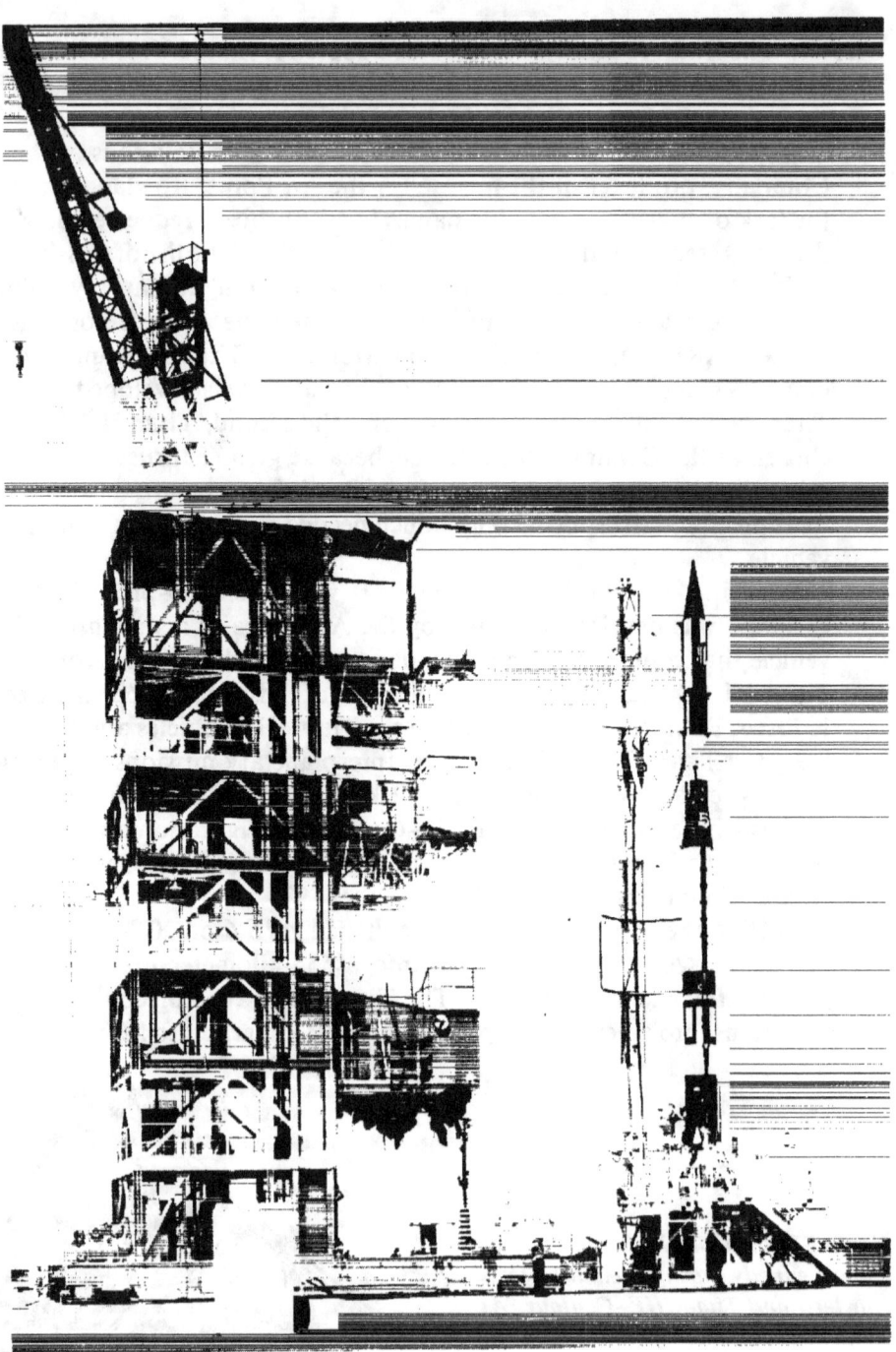

Vanguard

LAUNCH VEHICLES

VANGUARD. The name "Vanguard," adopted in 1955, applied to the U.S. International Geophysical Year satellite project as well as to the launch vehicle developed to orbit the satellites (see Vanguard under Satellites). Stages of the Vanguard rocket were later adapted to the NASA Delta vehicle.

II

SATELLITES

Full-disc photograph of the earth from equatorial orbit, transmitted by ATS 3 on 10 November 1967. A cold front moves eastward over the central United States and a tropical storm is at bottom center.

SATELLITES

Astronomy's traditional definition of a satellite is "a celestial body orbiting another of larger size." [1] Through the balance of gravitational attraction, velocity, and centrifugal force, the moon revolves around, or orbits, the earth; hence, it is a satellite of the earth. Since 1957, man has been using rocket-powered launch vehicles to place man-made objects in orbit around the earth. Because they orbit the earth, these objects are earth "satellites."

Technically, of course, orbiting manned spacecraft also became satellites of the earth. Other satellites, in the strict sense of the word, were the spent rocket stages and uninstrumented pieces of hardware—popularly called "space junk"—placed in orbit incidentally. For purposes of the space exploration program, the term "satellite" was applied to man-made, instrumented objects placed intentionally in earth orbit to perform specific functions.

NASA unmanned satellites are divided into two categories: scientific satellites (which obtain scientific information about the space environment) and applications satellites (which perform experiments that will have everyday usefulness for man on earth). Satellites in the Explorer series were typical of the scientific satellites, gathering a variety of scientific data and telemetering it to stations on earth. Examples of applications satellites were Tiros meteorological satellites, designed to provide cloud-cover photographs to aid in forecasting weather conditions, and Relay and later communications satellites, designed to receive and transmit voice and facsimile communications between distant points on the earth.

Artist's concept of Aeros in orbit.

AEROS. In June 1969 NASA and the German Ministry for Scientific Research (BMwF) reached an agreement on a cooperative project that would orbit a German scientific satellite designed to investigate particle behavior in the earth's upper atmosphere.[1] In early 1969 BMwF had named the proposed aeronomy satellite after Aeros, ancient Greek god of the air.[2] *Aeros,* the second U.S.–German cooperative research satellite, was designated GRS–A–2 by NASA (see also Azur) before launch, assuming its proper name when successfully launched into orbit 16 December 1972.[3] *Aeros 2* was orbited by NASA 16 July 1974.

("Aeros" also had been used earlier for the Synchronous Meteorological Satellite project [see SMS].)

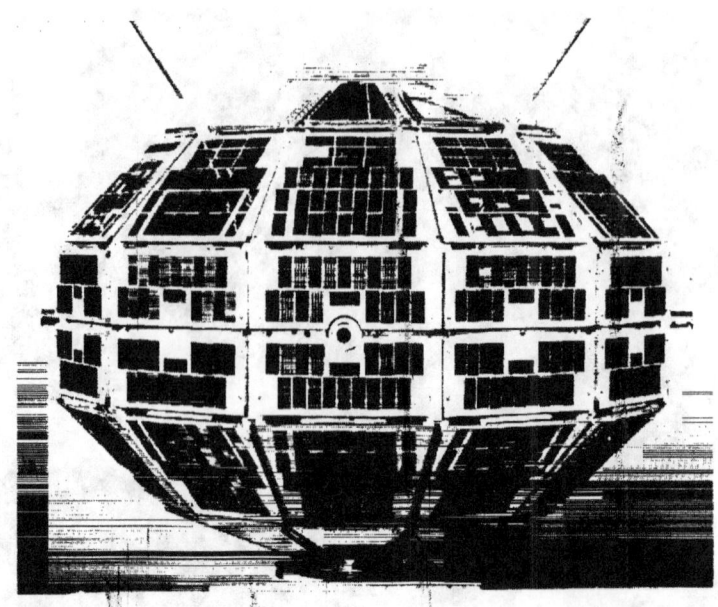

Alouette 1, *above, before launch, and an artist's concept of* ANS *below. Ariel 4, at right, was the first satellite in the Ariel series to carry a United States experiment.*

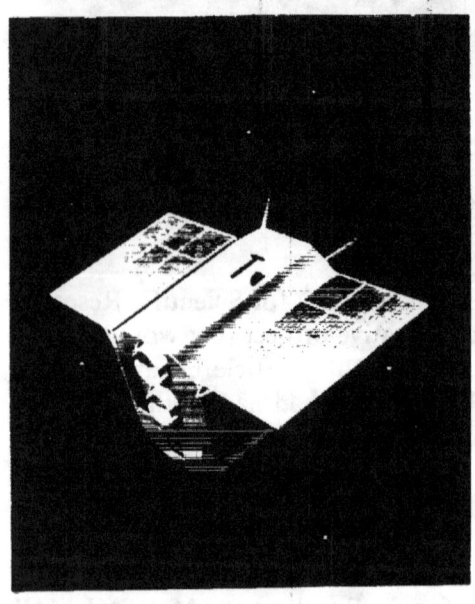

SATELLITES

ALOUETTE. An international satellite program, Alouette was a Canadian project in cooperation with NASA [1] and was given its name in May 1961 by the Canadian Defence Research Board. The name was selected because, as the French-Canadian name for meadowlark, it suggested flight; the word "Alouette" was a popularly used and widely known Canadian title; and, in a bilingual country, it called attention to the French part of Canada's heritage. [2] NASA supported the Board's choice of name for the topside sounder scientific satellite. [3]

Alouette 1, instrumented to investigate the earth's ionosphere from beyond the ionospheric layer, was launched into orbit by NASA from the Pacific Missile Range 28 September 1962. It was the first satellite designed and built by a country other than the United States or the Soviet Union and was the first satellite launched by NASA from the West Coast. *Alouette 2* was orbited later as part of the U.S.–Canadian ISIS project (see ISIS).

NASA's *Explorer 20,* launched 25 August 1964, was nicknamed "Topsi" for "topside sounder"; it returned data on the ionosphere to be compared with Alouette data, as well as data from *Ariel 1* and *Explorer 8* and sounding rockets. [4]

ANS. In June 1970 NASA and the Netherlands Ministries of Economic Affairs and Education and Science reached agreement to launch the first Netherlands scientific satellite in 1974. The satellite was designated "ANS," an acronym for "Astronomical Netherlands Satellite," and an ANS Program Authority was created by the Ministries to direct the cooperative project. NASA provided an experiment and the Scout launch vehicle, and the Program Authority designed, built, and tested the spacecraft and provided tracking and data acquisition. [1] The satellite—launched 30 August 1974— carried an ultraviolet telescope to study selected stellar ultraviolet sources and instruments to investigate both soft and hard x-ray sources.

ARIEL. The world's first international satellite, *Ariel 1* was a cooperative project between the United Kingdom and NASA. The satellite was named in February 1962 for the spirit of the air who was released by Prospero in Shakespeare's play *The Tempest.* The name "Ariel"—a traditional name in British aeronautics—was chosen by the U.K. Minister of Science and endorsed by NASA. [1] Other satellites followed in the program.

Ariel 1 (UK–1 before orbit), launched from Cape Canaveral 26 April 1962, was built by NASA's Goddard Space Flight Center and instrumented with six British experiments to make integrated measurements in the ionosphere. *Ariel 2*, containing three U.K.-built experiments, was placed in orbit 27 March 1964. *Ariel 3*, designed and built in the United Kingdom, was launched 5 May 1967 with five experiments. The U.K.-built *Ariel 4* carried four U.K. and one U.S. experiment into orbit 11 December 1971 to investigate plasma, charged particles, and electromagnetic waves in the ionosphere. *Ariel 5* (UK–5), also British-built, was launched 15 October 1974 to study x-ray sources.

The UK–X4 satellite was in a different series from the Ariels. An "X" added to the prelaunch designation indicated it was experimental and, when orbited March 1974 to test spacecraft systems and sensors, the spacecraft was christened *Miranda*. It was a United Kingdom satellite launched by NASA under a contract for reimbursable services, rather than a joint research mission.[2]

The United Kingdom's Skynet satellites belonged to still another series. The Skynet I and II series of U.K. Ministry of Defence communications satellites were launched by NASA, beginning in 1969, under agreement with the U.S. Air Force, which reimbursed NASA for launch vehicles and services.

ATS. The name "ATS"—an acronym for "Applications Technology Satellite"—referred to the satellite mission: to test technological experiments and techniques for new practical applications of earth satellites. The name evolved through several transitions, beginning with the project's study phase. In 1962–1963, at NASA's request, Hughes Aircraft Company conducted feasibility and preliminary design studies for an "Advanced Syncom" satellite. The concept was of a communications spacecraft in synchronous orbit with a new stabilization system and a multiple-access communication capability. Other names in use were "Advanced Synchronous Orbit Satellite," "Advanced Synchronous Satellite," and "advanced synchronous communications satellite."[1]

By March 1964 NASA had decided Advanced Syncom should not only test communications technology but also support development of "meteorological sensing elements, measurements of the space environment in various orbits such as the synchronous orbit, and the conduct of experiments on general stabilization systems which apply not only to communications systems but to other systems."[2] As the concept of the satellite was changed, so was its name—becoming "Advanced Technological Satellite (ATS)." Hughes was selected to build five ATS spacecraft.[3]

The change to "Applications Technology Satellite" came in October 1964. Dr. Homer E. Newell, NASA Associate Administrator for Space Science and Applications, and Dr. John F. Clark, Director of Space Sciences, had concluded that the adjectives "Advanced Technological" were undesirable because they seemed to conflict with responsibilities of NASA's Office of Advanced Research and Technology.[4] On 2 October Dr. Newell formally proposed, and Associate Administrator Robert C. Seamans, Jr., approved, the change to "Applications Technology Satellite"—bringing "the name of the project more into line with its purpose, applications technology, while retaining the initials ATS by which it is commonly known." [5]

Launched 6 December 1966, *ATS 1* took the first U.S. high-quality photographs of the earth from synchronous orbit, showing the changing cloud-cover patterns. In addition to weather data, the satellite relayed color television across the United States and voice signals from the ground to aircraft in flight. *ATS 3,* launched 5 November 1967, carried advanced communications, meteorology, and navigation experiments and made high-resolution color photographs of one complete side of the earth. *ATS 6* was launched 30 May 1974 to support public health and education experiments in the United States and India. It was the first communications satellite with the power to broadcast TV photos to small local receivers.

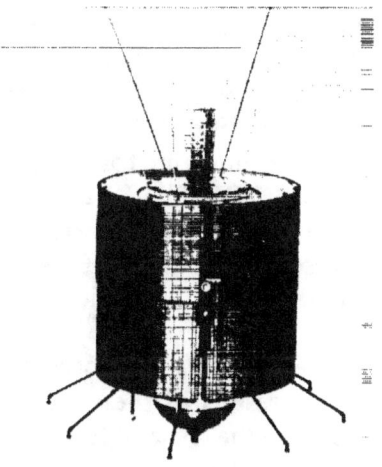

ATS 3, *at left, suspended during antenna pattern testing. The model of* ATS 6, *below, shows the nine-meter reflector deployed.*

AZUR. A 17 July 1965 memorandum of understanding between NASA and the German Ministry for Scientific Research (BMwF) initiated a cooperative project that would orbit a German scientific satellite to investigate the earth's inner radiation belt. The agreement provided for the launch of the satellite after a successful series of sounding rocket tests to check out the proposed satellite instrumentation.[1] NASA would provide the Scout launch vehicle, conduct launch operations, provide tracking and data acquisition, and train BMwF personnel. In June 1966 NASA designated the satellite GRS–A, an acronym for "German Research Satellite–A."* "Azur," the German word for the color "sky blue," was chosen by BMwF in early 1968 as the name for the satellite, and GRS–A was officially designated *Azur* by NASA after launch 7 November 1969.[2] (See also Aeros.)

BIOSATELLITE. As the name suggests, Biosatellites were used to conduct space experiments with living organisms, both plant and animal. The biological specimens in orbit underwent prolonged weightlessness, radiation, and other conditions of the space environment; scientists could study the effects on various life processes. Physiological effects included growth and form of entire organisms, structure of growth of cells and tissues, and basic biochemistry of the cell.[1]

The NASA Project Designation Committee, asked by the Director of Bioscience Programs in June 1962 to consider an official name for such a project should it be initiated, devised the name "Biosatellite," a contraction of the phrase "biological satellite." The shorter "Bios" formed the basis for the new name and occasionally appeared as a substitute for Biosatellite.[2] But Biosatellite should not be confused with "BIOS" ("Biological Investigation of Space"), the name of a separate reentry spacecraft flown in 1961. The Project Designation Committee reserved the name "Biosatellite" for project use, pending approval of the orbiting biological payload project.[3]

In March 1963 NASA contracted for spacecraft feasibility studies for a "bio-satellite program." After evaluating the results of these studies and obtaining funding for the project, NASA selected the General Electric Company to build the spacecraft and later chose the biological experiments to be flown on them. By early 1964 the project was well under way and the name "Biosatellite" had been adopted.

---

*"GRS–A" became "GRS–A–1" when an agreement was reached to orbit a second research satellite, designated "GRS–A–2" (Aeros). (NASA, "Project Approval Document," 15 June 1966.)

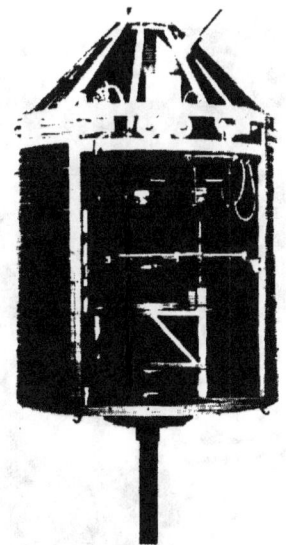

Cutaway models of the German-built Azur, *at left, and NASA's Biosatellite, below.*

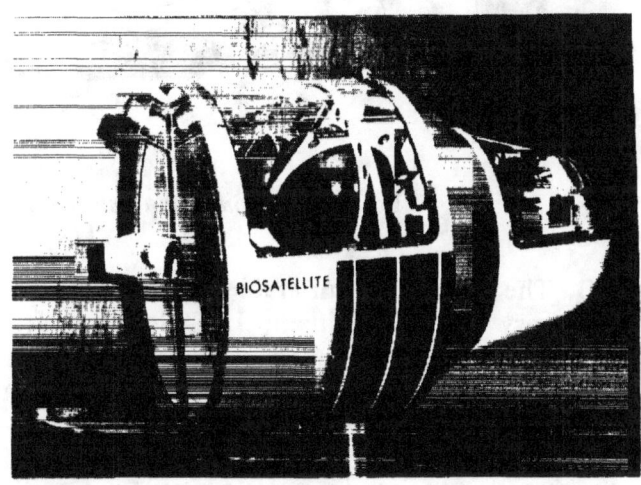

Biosatellite 1 was launched 14 December 1966; it functioned normally in orbital space flight but failed to reenter as it should have three days later. *Biosatellite 2,* launched 7 September 1967, obtained information on the effects of radiation and weightlessness on plant and low-order animal life forms. The program ended with the flight of *Biosatellite 3,* launched 28 June 1969, which was prematurely terminated after eight and one-half days. Analysis of the death of the pigtailed monkey orbited during that flight provided additional information on the effects of prolonged weightlessness during manned flights.

Echo 1 *during inflation tests before launch.*

ECHO. The idea of an inflatable, spherical space satellite was conceived in
January 1956 by William J. O'Sullivan, Jr., aeronautical engineer at
NACA's Langley Aeronautical Laboratory (later NASA Langley Research
Center), and proposed as an air-density experiment for the International
Geophysical Year (1 July 1957 to 31 December 1958).[1] The balloon satellite
was similar to one described by John R. Pierce of Bell Telephone
Laboratories in his 1955 article, "Orbital Radio Relays."[2] Pierce was in-
terested in the orbiting inflated sphere for use as a reflector for radio signals
and he proposed a cooperative communication experiment using
O'Sullivan's balloon satellite. By early 1959 O'Sullivan's original proposal
for IGY air-density studies had become NASA's passive communications
satellite project.[3]

The word "echo" was often used in the radio and radar sense to describe
the reflection of ground-transmitted signals from the surface of an orbiting
balloon. The name "Project Echo," derived through informal use, was given
to the 30-meter inflatable-structure satellite.[4] O'Sullivan's design was tested
in a series of Shotput launches and the Echo project proved that an

aluminized-Mylar sphere could be carried aloft by a rocket, be inflated in space, and remain in orbit to provide a means of measuring atmospheric density as well as a surface for reflecting radio communications between distant points on the earth.

Echo 1 passive communications satellite, orbited by NASA 12 August 1960, was the fruition of O'Sullivan's labors. His inflatable-sphere concept also was employed in three air-density Explorer satellites, in *Echo 2,* and in *Pageos 1.*

EOLE. NASA and France's Centre National d'Études Spatiales (CNES) signed a memorandum of understanding 27 May 1966 providing for development of a cooperative satelite-and-instrumented-balloon network to collect meteorological data for long-range weather forecasts.[1] "Eole," the French name for Aeolus, ancient Greek god of the winds, was chosen by CNES as the name for the satellite project.[2] Known as "FR-2 [see also *FR-1*] until late 1968—and also as simply "French Satellite" before December 1968—the project was redesignated by NASA "CAS-A," an acronym for the first in a series of international "Cooperative Applications Satellite(s)."[3] The satellite was given its permanent name *Eole* after successful launch into orbit 16 August 1971.

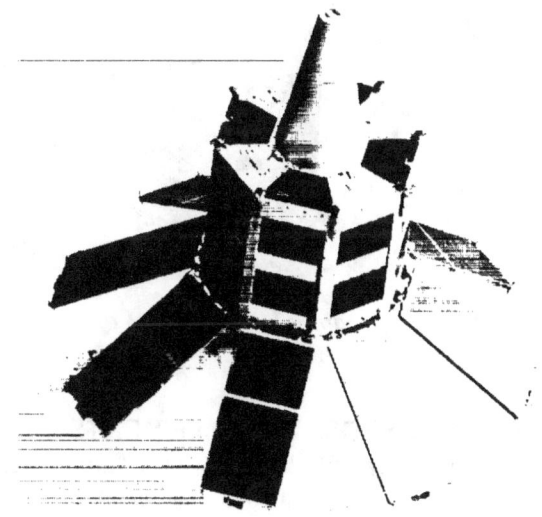

Eole, *cooperative French and U.S. satellite to collect meteorological data for forecasts.*

ERTS 1 (*renamed* LANDSAT 1) *after prelaunch tests at General Electric.*

ERTS, EOS, SEOS. The name "ERTS"—an acronym for "Earth Resources Technology Satellite"—was a functional designation; it was derived from early concepts of an "earth resources" satellite system to provide information on the environment by using remote-sensing techniques. Between 1964 and 1966, studies of remote-sensing applications were conducted jointly by NASA and the Departments of Interior and Agriculture and NASA initiated a program of aircraft flights to define sensor systems for remote-sensing technology. The studies indicated that an automated remote-sensing satellite appeared feasible and that a program should be initiated for the development of an experimental satellite.[1]

In early 1967 NASA began definition studies for the proposed satellite, by then designated ERTS, and by early 1969 the project was approved.[2] Two satellites, ERTS–A and ERTS–B, were subsequently planned for launch. ERTS–A became *ERTS 1* on launch 23 July 1972; it was still transmitting data on earth resources, pollution, and environment at the end of 1974, for users worldwide. ERTS–B was scheduled for 1975 launch.[*]

The early nomenclature for both the program and the proposed satellites was confusing. The "Earth Resources Program" was variously known as the "Natural Resources Program," the "Earth Resources Survey Program," and the "Earth Resources Observation Program."[3] The designation "Earth Resources Survey Program" was eventually used to include ERTS and remote-sensing aircraft programs, as well as the "Earth Resources Experiment Package" (EREP) flown on Skylab missions in 1973–1974. These programs formed a part of NASA's overall "Earth Observations Programs," which also included the meteorology and earth physics program.[4]

Before 1967 several names were in use for the proposed earth resources satellite, including the designation "ERS"—a shortened acronym for "Earth Resources Survey Satellite"—which was in conflict with an identical designation for an Air Force satellite project known as the "Environmental Research Satellite."[5] Further confusion arose when the Department of the Interior, which in cooperation with NASA had been studying the application of remote-sensing techniques, announced the name "EROS"—an acronym for "Earth Resources Observation Satellite"—for the satellite

---

[*]NASA announced 14 January 1975 that *ERTS 1* had been renamed *LANDSAT 1* and ERTS–B would become *LANDSAT 2* when launched 22 January. Associate Administrator for Applications Charles W. Mathews said that, since NASA planned a SEASAT satellite to study the oceans, "LANDSAT" was an appropriate name for the satellite that studied the land. Dr. George M. Low, Deputy Administrator of NASA, had suggested a new name be found for ERTS, with more public appeal. John P. Donnelly, Assistant Administrator for Public Affairs, had therefore requested NASA office heads and Centers to submit ideas for new names by the end of December 1974. From a number of replies received, the NASA Project Designation Committee made its recommendation, and "LANDSAT" was approved. (NASA, News Release 75-15; Mathews, ERTS–B Mission Briefing, NASA Hq., 14 Jan. 1975; Howard G. Allaway, Public Affairs Officer, NASA, telephone interview, 3 Feb. 1975; and Bernice Taylor, Administrative Assistant to Assistant Administrator for Public Affairs, NASA, telephone interview, 12 Feb. 1975.)

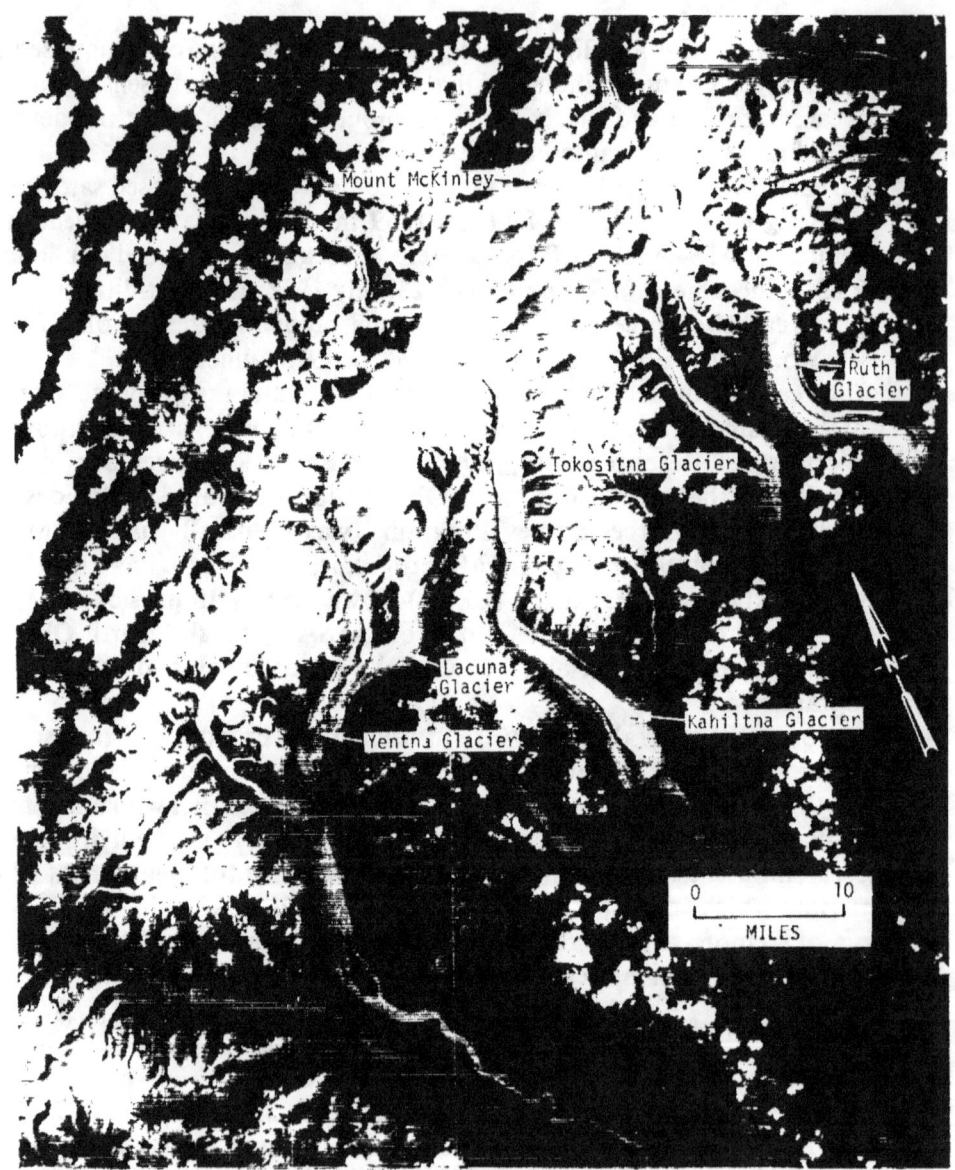

"Galloping," or surging, glaciers in Alaska identified by U.S. Geological Survey scientists on images taken by ERTS 1. Glacier behavior could give warning of floods and clues to sources of water supply.

project.[6] In early 1967, when NASA initiated the definition studies of the experimental satellite, the name "ERTS" came into use.[7]

In early 1970 the NASA Project Designation Committee met to choose a new name for the ERTS satellites and several names were suggested, including "Earth," "Survey," and "Ceres"—the ancient Greek goddess of the harvest. The Committee favored "Earth" but, after submitting the name to the other Government agencies in the program and receiving unfavorable responses from some, it dropped the name "Earth," and "ERTS" was used up through the end of 1974.[8]

EOS. A follow-on to ERTS was to be the Earth Observatory Satellite (EOS)—given a functional name in NASA. Results from the first Earth Resources Technology Satellite showed that greater spectral and spatial resolutions were needed for some applications, such as classifying and monitoring the use of land for urban planners and increasing accuracy in predicting agricultural yield. And since 1970 NASA had seen a need for a multipurpose satellite in low earth orbit to survey the earth and oceans, detect pollution, and monitor the weather. Definition studies were begun in 1974 of a low-cost EOS spacecraft that could be launched, resupplied or serviced, and eventually returned by the Space Shuttle, but could also be launched by a conventional booster before the shuttle became operational. Modular systems for power and different spacecraft functions would permit the Shuttle to unplug and replace malfunctioning systems.[9]

EOS–A was tentatively scheduled for 1979 launch as a land-and-water-use mission, with EOS–B possibly in 1981.

SEOS. An advanced study also was under way in 1974 of a Synchronous Earth Observatory Satellite (SEOS) for experimental meteorological and earth resources observations using a large telescope with improved resolution and an infrared atmospheric sounder. The geosynchronous orbit would provide the short intervals needed to detect and warn of natural disasters such as hurricanes, tornadoes, forest fires, floods, and insect crop damage.[10]

ESRO. The European Space Research Organization (ESRO), a 10-member Western European group to conduct scientific space research, came into formal existence in March 1964 (the ESRO Convention had been signed 14 June 1962). The Organization named its first satellites "ESRO" in honor of its own abbreviation.[1] An 8 July 1964 NASA–ESRO agreement originally called for two cooperative satellites, ESRO 1 to investigate the polar ionosphere and ESRO 2 to study solar astronomy and cosmic rays. With development of the scientific payloads, it became apparent that ESRO 1 had a rather narrow launch opportunity and that it was important to launch it in

European designed and built ESRO 1 *was also given the name* Aurorae *in orbit. In the photo the satellite was being tested at the Western Test Range before launch.*

the fall; therefore ESRO 2 was moved up for first launch, although the number designations were not changed.[2]

After launch by NASA on 3 October 1968, *ESRO 1* was also assigned the name *Aurorae* by ESRO; it was designed to study the aurora borealis and related phenomena of the polar ionosphere. Its numerical designation later became *ESRO 1A* when a duplicate backup satellite, *ESRO 1B,* was launched 1 October 1969. *ESRO 1B* was designated *Boreas* by ESRO.[3]

ESRO 2A, scheduled to be the first ESRO satellite, failed to reach orbit 29 May 1967. Its backup, *ESRO 2B,* was given the name *IRIS*—an acronym for

SATELLITES

"International Radiation Investigation Satellite"—by ESRO after successful launch 16 May 1968.[4]

Under the 1964 memorandum of understanding, NASA's participation in the cooperative venture was to provide Scout launch vehicles, conduct launch operations, provide supplemental tracking and data acquisition services, and train ESRO personnel. No funds were exchanged in the project.[5]

Under a 30 December 1966 memorandum of understanding, ESRO became the first international space group to agree to pay NASA for launchings; it would reimburse NASA for launch vehicle and direct costs of equipment and services. The first satellite orbited under this agreement, *HEOS 1*—"Highly Eccentric Orbit Satellite"—was launched 5 December 1968.[6]

Later scientific and applications satellites planned by ESRO—and to be launched by NASA—were given functional names:[7]

> Cos-B, scheduled for 1975 launch to study cosmic rays from the galaxy, especially gamma radiation, was to be one of the successors to the *TD-1A* astronomical satellite launched by NASA for ESRO in March 1972.
>
> GEOS, "Geostationary Scientific Satellite" (a different satellite from NASA's Geodetic Explorers or Geodynamic Experimental Ocean Satellite), was scheduled for 1976 launch to study cosmic radiation over a long period.
>
> EXOSAT, a high-energy astronomy satellite, was planned for 1979 launch for x-ray astronomy.
>
> METEOSAT, a geostationary meteorological satellite, was planned for 1976 launch.
>
> OTS, geostationary "Orbital Test Satellite," was to be launched in 1976 or 1977 as a forerunner of the European Communications Satellite (ECS; formerly CEPT, for Conférence Européene des Postes et Télécommunications, or CETS, for Conference on European Telecommunications Satellite).
>
> AEROSAT, joint "Aeronautical Satellite" to be developed with the U.S. Federal Aviation Agency and a U.S. contractor, was to be launched in 1977 or 1978 for air traffic control, navigation, and communications.
>
> MAROTS, "Maritime Orbital Test Satellite," an adaptation of OTS funded principally by the United Kingdom, was planned for 1977 launch for civil maritime communications and navigation.

ESRO was also cooperating with NASA in the International Sun-Earth Explorer (ISEE) program and the International Ultraviolet Explorer (IUE) program (see Explorer).

Engineers check out ESSA 5 in the top photo. The first complete coverage of North America from ESSA 1, in the lower photo, was transmitted 5 February 1966.

ESSA. The ESSA satellites were meteorological satellites in the Tiros Operational Satellite (TOS) system that were financed and operated by the Environmental Science Services Administration (ESSA). The name was selected by ESSA early in 1966 and was an acronym derived from "Environmental Survey Satellite"; [1] it was also the abbreviation for the operating agency.

Between 1966 and 1969 NASA procured, launched, and checked out in orbit the nine ESSA satellites, beginning with *ESSA 1,* orbited 3 February 1966.

On 3 October 1970 ESSA was incorporated into the new National Oceanic and Atmospheric Administration (NOAA). After launch by NASA, the subsequent series of satellites—in the Improved TOS (ITOS) system—were turned over to NOAA for operational use. The first ITOS spacecraft funded by NOAA, launched 11 December 1970, was designated *NOAA 1* in orbit,[2] following the pattern set by the ESSA series. (See also Tiros, TOS, and ITOS.)

EXPLORER. The name "Explorer," designating NASA's scientific satellite series, originated before NASA was formed. "Explorer" was used in the 1930s for the U.S. Army Air Service-National Geographic stratosphere balloons. On 31 January 1958, when the first U.S. satellite was orbited by the U.S. Army as a contribution to the International Geophysical Year (IGY), Secretary of the Army Wilbur M. Brucker announced the satellite's name, *Explorer 1*. The name indicated the mission of this first satellite and its NASA successors—to explore the unknown.[1]

The Army Ballistic Missile Agency (ABMA) had previously rejected a list of explorer names for the satellite. Jet Propulsion Laboratory, responsible for the fourth stage of the Jupiter C rocket (configured as the Juno I launch vehicle) and for the satellite, had called the effort "Project Deal" (a loser in a poker game always called for a new deal—and this satellite was the answer to the Russian Sputnik). On the day of the launch, ABMA proposed the name "Top Kick," which was not considered appropriate. The list of names was brought out again. All the names on the list had been crossed out and only the heading "Explorers" remained. The late Richard Hirsch, a member of the National Security Council's Ad Hoc Committee for Outer Space, suggested that the first American satellite be called simply "Explorer." The name was accepted and announced.[2]

When NASA was being formed in 1958 to conduct the U.S. civilian space program, responsibility for IGY scientific satellite programs was assigned to NASA. The decision was made by the National Advisory Committee for Aeronautics (NACA) to continue the name "Explorer" as a generic term for future NASA scientific satellites.[3] Explorers were used by NASA to study (1) the atmosphere and ionosphere, (2) the magnetosphere and interplanetary space, (3) astronomical and astrophysical phenomena, and (4) the earth's shape, magnetic field, and surface.

Many of the Explorer satellites had project names that were used before they were orbited and then supplanted by Explorer designations once they were placed in orbit. Other Explorer satellites, particularly the early ones, were known before orbit simply by numerical designations. A listing of some of the Explorers' descriptive designations illustrates the variety of scientific missions performed by these satellites: Aeronomy Explorer, Air

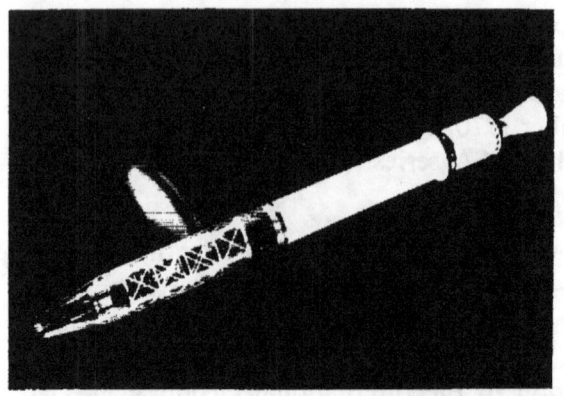

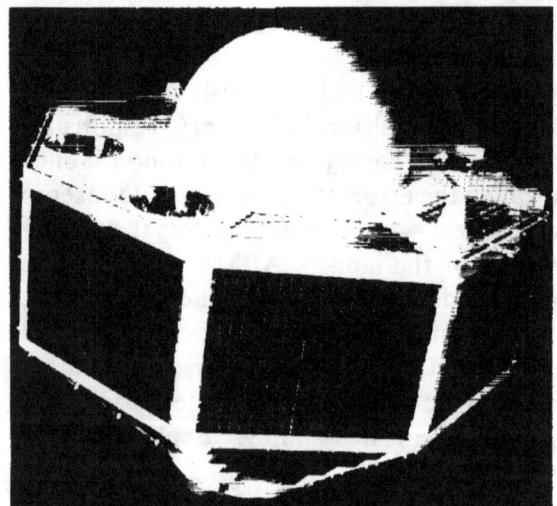

Some of the Explorer satellites, left to right above: Explorer 1; Explorer 11 *Gamma-ray Astronomy Satellite; and* Explorer 29 *Geodetic Satellite.*

Density Satellite, Direct Measurement Explorer, Interplanetary Monitoring Platform (IMP), Ionosphere Explorer, Meteoroid Technology Satellite (MTS), Radio Astronomy Explorer (RAE), Solar Explorer, Small Astronomy Satellite (SAS).

SAS–A, an X-ray Astronomy Explorer, became *Explorer 42* when launched 12 December 1970 by an Italian crew from the San Marco platform off the coast of Kenya, Africa. It was also christened *Uhuru,* Swahili for "Freedom," because it was launched on Kenya's Independence Day. The small satellite, mapping the universe in x-ray wavelengths for four years, discovered x-ray pulsars and evidence of black holes.[4]

Geodetic Satellites (GEOS) were also called "Geodetic Explorer Satellites" and sometimes "Geodetic Earth Orbiting Satellites." *GEOS 1 (Ex-*

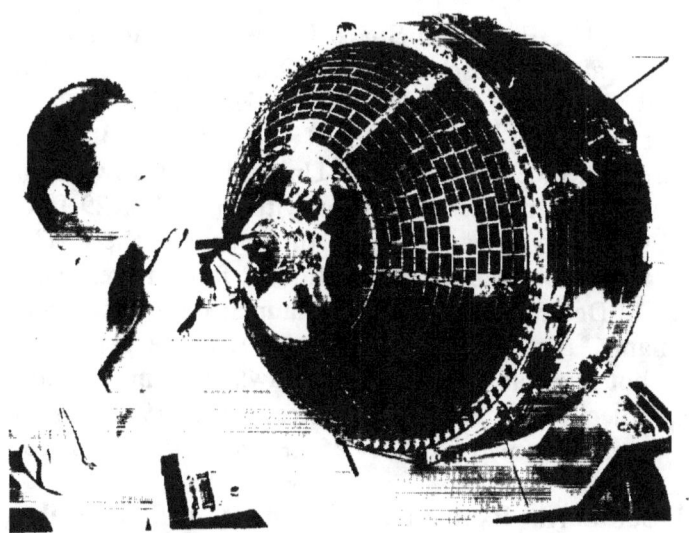

Atmosphere Explorer (above), named Explorer 32 *in orbit, and an artist's drawing of an Interplanetary Monitoring Platform* (Explorer 34) *at right.*

plorer 29, launched 6 November 1965) and *GEOS 2* (*Explorer 36,* launched 11 January 1968) refined knowledge of the earth's shape and gravity field. GEOS–C, to be launched in 1975 as a successor to *GEOS 1* and *2,* was renamed "Geodynamic Experimental Ocean Satellite" to emphasize its specific mission in NASA's earth and ocean physics program while retaining the GEOS acronym. GEOS–C was to measure ocean currents, tides, and wave heights to improve the geodetic model of the earth and knowledge of earth-sea interactions. (The European Space Research Organization's Geostationary Scientific Satellite—also called GEOS, planned for 1976 launch—was not a part of the Geodetic Explorer series. See ESRO.) [5]

The 52nd Explorer satellite was launched by NASA 3 June 1974—*Hawkeye 1*, also called *Explorer 52*, a University of Iowa-built spacecraft. The University's Injun series had begun with *Injun 1* on 29 June 1961, to study charged particles trapped in the earth's magnetosphere. The first three Injuns were launched by the Air Force (Injun 2 failed to reach orbit; *Injun 3* was orbited 13 December 1962). NASA launched the next three, adding the Explorer name. *Hawkeye 1* originally carried the prelaunch designation "Injun F" but this was discarded; the Hawkeye name was approved by the NASA Project Designation Committee in June 1972. (*Injun 4*, 21 November 1964, was also named *Explorer 25; Injun 5*, 8 August 1968, was *Explorer 40*.) [6]

Two International Sun-Earth Explorers, ISEE-A (sometimes called "Mother") and ISEE-B (sometimes called "Daughter"), were planned for dual launch in 1977, to be followed by ISEE-C ("Heliocentric") in 1978. The joint NASA and European Space Research Organization program—earlier called the International Magnetosphere Explorer (IME) program—was to investigate sun-earth relationships and solar phenomena. [7]

An International Ultraviolet Explorer (IUE; originally designated SAS-D in the Small Astronomy Satellite series) was scheduled for 1976 launch as a cooperative NASA, United Kingdom, and ESRO satellite to gather high-resolution ultraviolet data on astronomical objects. [8]

An Applications Explorer, the Heat Capacity Mapping Mission (HCMM) was planned for 1977 launch. A "small, dedicated satellite," the HCMM was to be a simple, low-cost spacecraft with one sensor for one purpose, making thermal measurements of the earth's surface across the United States. Measurements would map kinds of rocks and soil, help find mineral resources, and show whether geothermal energy sources could be found by spacecraft. [9]

FR-1. *FR-1* was the designation of the French satellite orbited by NASA 6 December 1965 in a cooperative U.S.-French program to investigate very-low-frequency electromagnetic waves. The name developed in 1964, when NASA and France's Centre National d'Études Spatiales (CNES) agreed, after preliminary sounding rocket experiments, to proceed with the satellite project. [1] CNES provided the satellite and designated it "FR-1" for "France" or "French" satellite number one. [2] The first flight unit was

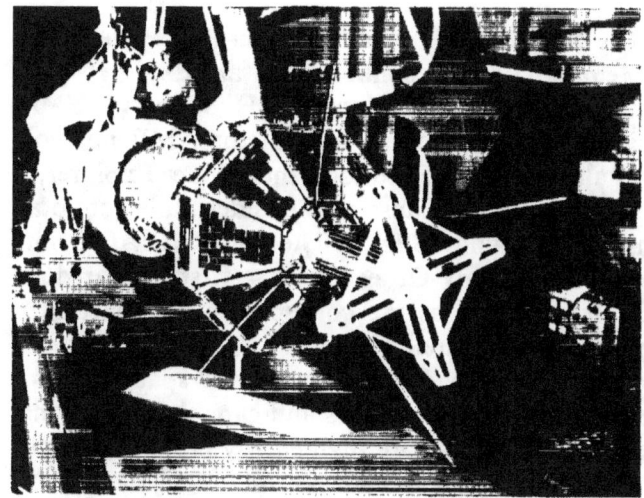

FR–1 *in testing at the Western Test Range.*

Artist's concept of HEAO.

designated "FR–IA" and the backup unit, "FR–IB." The second U.S.–France cooperative satellite, "FR–2," was later renamed "Eole." (See Eole.)

HEAO. In September 1967 NASA established the Astronomy Missions Board to consult the scientific community and submit for consideration a long-range program for the 1970s. The Board's X-Ray and Gamma-Ray Panel completed its report in September 1968, recommending an Explorer-class spacecraft with a larger payload capability, designated "High Energy A" by the Panel and "Heavy Explorer" in other sections of the AMB position paper.[1] The spacecraft was alternately referred to as the "Super Explorer," but all three names were later dropped because of the undesirable connotation of their abbreviations ("HEX" and "SEX"). The name "HEAO"—an acronym for "High Energy Astronomy Observatory"—first appeared in June 1969 and was officially adopted as the concept for the spacecraft evolved to that of an observatory-class satellite.[2]

HEAO was originally planned to be the largest unmanned spacecraft orbited by the U.S., weighing almost 10 metric tons and capable of carrying the larger instruments required to investigate high-energy electromagnetic radiation from space—including x-rays, gamma rays, and high energy cosmic rays. The first satellite in the series, HEAO–A, was to be launched by a Titan IIIE launch vehicle in 1975.

In January 1973 the project was suspended because of budget cuts. A scaled-down project was substituted in FY 1975, calling for three spacecraft instead of four, to be launched by an Atlas-Centaur vehicle instead of the Titan IIIE, in 1977, 1978, and 1979. With the smaller launch vehicle, HEAO was revised to carry fewer instruments and weigh about 3200 kilograms. The first mission was to make an x-ray survey, the second detailed x-ray studies, and the third a gamma and cosmic ray survey of the sky. Launches of spacecraft from NASA's Space Shuttle after 1980 would carry heavier gamma and cosmic ray experiments to complete the scientific objectives.[3]

HEOS. The name of the HEOS satellite, built and named by the European Space Research Organization (ESRO), is an acronym for "Highly Eccentric Orbit Satellite."[1] *HEOS 1* was launched 5 December 1968 to investigate interplanetary magnetic fields and study solar and cosmic ray particles outside the magnetosphere. Nine scientific groups in five countries provided experiments on board the satellite. Under a 30 December 1966 memorandum of understanding and an 8 March 1967 contract with ESRO, the mission was the first cost-reimbursed NASA launch of a foreign scientific satellite.[2] *HEOS 2*, the second satellite in the series, was launched by NASA 31 January 1972, to continue the study of the interplanetary medium. (See also ESRO.)

INTASAT. NASA and Spain signed a memorandum of understanding in May 1972 on a joint research program in which NASA would launch Spain's first

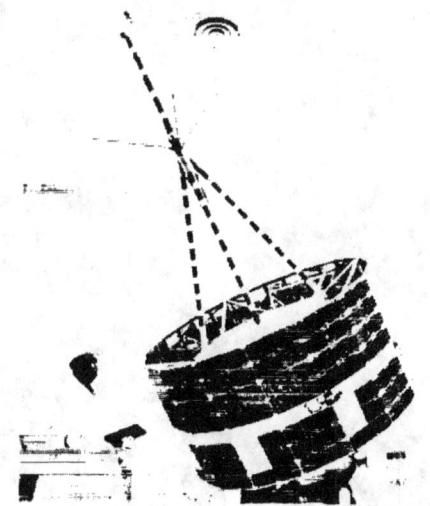

HEOS 1, *at left, in preparation for its 30 December 1966 launch with experiments from five countries.* INTASAT, *below, was launched as Spain's first satellite 15 November 1974.*

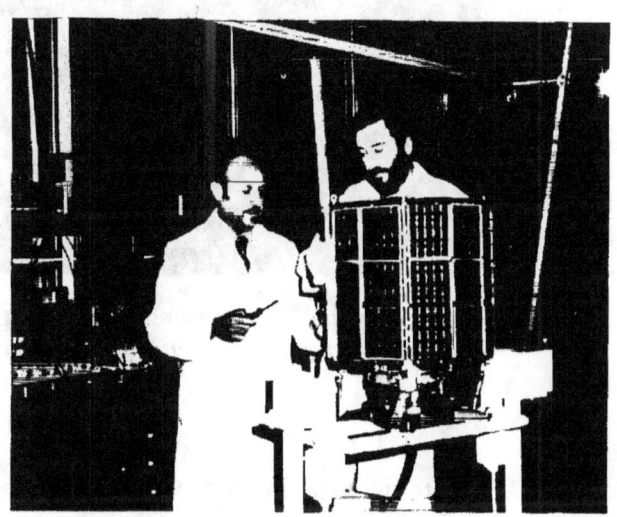

satellite. The Spanish Space Commission—Comision Nacional de Investigación del Espacio (CONIE)—named the satellite "INTASAT," an acronym for the Instituto Nacional de Téchnica Aeroespacial (INTA), the government laboratory responsible for development of the satellite. Designed and developed in Spain to measure the total electron count in the ionosphere and ionospheric irregularities, *INTASAT* was launched pickaback in a three-satellite launch (with *NOAA 4* and *OSCAR 7*) on 15 November 1974. The 15-kilogram satellite was to beam data to 25 to 30 scientists around the world from its sun-synchronous, polar orbit for two years.[1]

An Intelsat IV series comsat jettisons its nosecone shroud in an artist's drawing. The Centaur stage of the launch vehicle is still attached.

INTELSAT. Intelsat satellites were owned and operated by the International Telecommunications Satellite Organization (INTELSAT). They were launched and tracked, on a reimbursable basis, by NASA for the Communications Satellite Corporation, the U.S. representative in and manager of INTELSAT. INTELSAT's method of designating its satellites went through numerous changes as new satellites were launched, producing alternate names for the same satellite and varying the numbering system.

The first of the INTELSAT satellites, *Intelsat I,* was named "Early Bird" because it was the satellite in the "early capability program"—the program to obtain information applicable to selection and design of a global commercial system and to provide experience in conducting communications satellite operations.[1] *Early Bird,* the world's first commercial comsat, was launched by NASA 6 April 1965 and placed in synchronous orbit over the Atlantic Ocean.

Intelsat II–A, also called "Lani Bird," was the first communications satellite of the Consortium's Intelsat II series. *Lani Bird* was launched in October 1966 to transmit transpacific communications, but failed to achieve synchronous orbit. It was named by the Hawaiian press; "Lani" meant "bird of heaven." [2] *Intelsat II–B,* or *Pacific 1,* the second in the Intelsat II series, was launched in January 1967 and placed in orbit to provide transpacific service.[3] *Intelsat II–C* (later redesignated *Intelsat-II F–3* for flight 3 in series II), or *Atlantic 2,** was the second INTELSAT satellite to provide transatlantic service.[4] It was placed in synchronous orbit over the Atlantic in March 1967.

Subsequent satellites followed the same sequences: *Intelsat II–D,* or *Pacific 2,* was launched in September 1967 and later renumbered *Intelsat-II F–4;* Intelsat III–A (later Intelsat-III F–1) failed to achieve orbit in September 1968; *Intelsat-III F–2,* or *Atlantic 3,* was launched in December 1968.[5]

Satellites in the Intelsat IV series were numbered according to a different system, beginning with *Intelsat-IV F–2,* launched 25 January 1971. Although *Intelsat-IV F–2* was the first in the series to be launched, the "F–2" referred to the second "fabrication"—the second satellite built—rather than the second "flight" in the series.[6] Other satellites in the series followed this pattern, with *Intelsat-IV F–8* launched into orbit 21 November 1974.

Each successive series of satellites increased in size and communications capacity: satellites in the Intelsat II series were improved versions of *Early Bird;* Intelsat III satellites had 5 times the communications capacity of the II series; and Intelsat IV satellites not only had an increased capacity—more than 5 times that of the III series—but also were nearly 10 times as heavy.

IRIS. "IRIS," an acronym for "International Radiation Investigation Satellite," was designed, developed, and built by the European Space Research Organization (ESRO). ESRO assigned the name to *ESTRO 2B*—a backup satellite to ESRO 2A, which had been launched 29 May 1967 but had failed to achieve orbit.[1] Under an agreement with ESRO, NASA

---

*UPI nicknamed it "Canary Bird" because of the association with the Canary Islands earth station. "Canary Bird" appeared widely in the press as its designation, but was not adopted by INTELSAT.

launched *IRIS 1* on 16 May 1968, to study solar astronomy and cosmic ray particles. (See also ESRO.)

(NASA also briefly used a sounding rocket with the name "Iris.")

ISIS. ISIS was a cooperative satellite project of NASA and the Canadian Defence Research Board to continue and expand ionospheric experiments of the *Alouette 1* topside sounder satellite. The name was devised in January 1963 by John Chapman, project manager of the Canadian team; Dr. O. E. Anderson, NASA Office of International Affairs; and other members of the topside sounder Joint Working Group. They selected "Isis" because it was the name of an ancient Egyptian goddess and an acronym for "International Satellites for Ionospheric Studies." [1]

The first ISIS launch, known as "ISIS-X," was achieved 28 Nov. 1965, when NASA launched *Alouette 2* and *Explorer 31* from Western Test Range with a single Thor-Agena B booster. The Canadian topside sounder and the U.S. Direct Measurement Explorer were designed to complement each other's scientific data on the ionosphere. Both *ISIS 1* (launched 29 January 1969) and *ISIS 2* (launched 31 March 1971) carried experiments to continue the cooperative investigation of the ionosphere.

In 1969 the Canadian government proposed the substitution of an experimental communications satellite for the last of the projected ISIS spacecraft (ISIS-C). [2] The satellite was redesignated "CAS-C"—an acronym used by NASA to denote an international "Cooperative Applications Satellite." [3] In April 1971 a memorandum of understanding was signed by NASA and the Canadian Department of Communication providing for the launch of CAS-C, which later was again redesignated "CTS-A," an acronym for "Communications Technology Satellite." [4] CTS-A was scheduled for 1975 launch.

LAGEOS. In 1971 NASA was considering the possibility of launching a passive satellite, "Cannonball," on a Saturn launch vehicle left from the Apollo program. Definition and documentation were completed in late 1971. Subsequently the Office of Applications began defining a similar but less costly satellite as a new project to begin in Fiscal Year 1974. The redefined satellite was given the functional name "Laser Geodynamic Satellite," or "LAGEOS." LAGEOS was to be the first of a series of varied satellites within NASA's earth and ocean physics applications program (EOPAP)—including spacecraft launched on unmanned vehicles in 1976 and 1977 and later ones on the Space Shuttle. [1]

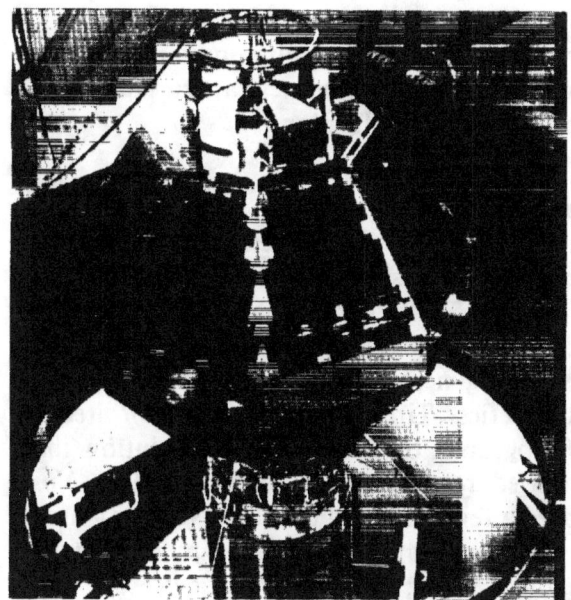

ISIS 2, *at left, carried 12 ionosphere experiments into orbit 31 March 1971. Below, technicians complete the final assembly of the LAGEOS satellite structure, which was to carry 426 precision optical laser retroreflectors into orbit in 1976.*

Approved as a "new start" for Fiscal Year 1974, with a 1976 launch date, the terrestrial reference satellite was to be a very heavy ball—weighing 411 kilograms although less than a meter in diameter—covered with laser reflectors to permit highly accurate measurements of the earth's rotational movements and movements of the earth's crust. The orbit and the weight of the simple, passive satellite were planned to provide a stable reference point for decades. The high, 5900-kilometer orbit would permit simultaneous measurements by laser ranging from earth stations a continent apart. Data would be used in earthquake prediction and other applications.[2]

NIMBUS. The meteorological satellite Nimbus was named from the meteorological term meaning "precipitating clouds" (from the Latin "rainstorm" or "cloud"). The satellite name was suggested in late 1959 by Edgar M. Cortright, Chief of NASA's Advanced Technology Programs, who directed the formation of NASA's meteorological satellite programs, including Nimbus and Tiros.[1] Nimbus was a second-generation research satellite following the first meteorological satellite series, Tiros. *Nimbus 1* was orbited 28 August 1964 and provided photographs of much higher resolution than those of Tiros satellites until it ceased transmission 23 September 1964. *Nimbus 2* (1966) and *3* (1969) operated a few years, followed by *Nimbus 4* and *5* in April 1970 and December 1972, to continue providing meteorological data and testing a variety of weather-sensing and measuring devices.

NOAA. From 1970, ITOS meteorological satellites launched by NASA were financed and operated by the National Oceanic and Atmospheric Administration (NOAA), which was established 3 October 1970 and incorporated the Environmental Science Services Administration (ESSA). Following ESSA's tradition of using the agency's acronym for the satellite name, the new series was named NOAA. *NOAA 1* (ITOS–A—following after the experimental Tiros-M, which had become *ITOS 1* on launch 23 January 1970) was launched 11 December 1970 to begin the new operational series. (See ESSA; and TIROS, TOS, and ITOS.)

OAO. The first satellite proposed for the "Orbiting Observatory" series, an astronomy satellite, was called the "Orbiting Astronomical Observatory" in early planning documents. It retained its original designation through the years, with the abbreviation OAO used as a short title. The term "Orbiting Astronomical Observatories" was first mentioned in writing by Dr. James E. Kupperian, Jr., in a December 1958 draft project outline, and NASA project officials approved this name as a working designation.[1] The question of a new name arose in March 1959 when NASA was preparing the first official project document. The long name had been shortened in common usage to "OAO." The NASA officials—Dr. Kupperian, Dr. G. F. Schilling, and Dr. Nancy Roman—decided to keep the long title with OAO as a short

SATELLITES

Nimbus 1, *at right, and the photograph it returned of Hurricane Alma on 11 September 1964. At bottom OAO 3 being checked out at Kennedy Space Center.*

title. The intent at the time was to keep a meaningful name, one which was short, descriptive, and professional.[2]

The first satellite of the program, *OAO 1*, was launched into almost perfect orbit 8 April 1966, but its power supply failed. *OAO 2*, launched 7 December 1968, took the first ultraviolet photographs of stars, returning data previously unobtainable. *OAO 3*, launched 21 August 1972, contained the largest telescope orbited by the U.S. to that date. It was given the additional name *Copernicus* after launch in honor of the Polish astronomer as part of the international celebration of the 500th anniversary of his birth.[3]

OFO, *the Orbiting Frog Otolith spacecraft, studied the adaptability of the inner-ear balance mechanism to weightlessness.*

OFO. "OFO" was an acronym for "Orbiting Frog Otolith"—not to be confused with similar acronyms describing the Orbiting Observatory series of spacecraft. The name, derived through common use, was a functional description of the biological experiment carried by the satellite ("otolith" referred to the frog's inner-ear balance mechanism).[1]

The Frog Otolith Experiment (FOE) was developed by Dr. Torquato Gualtierotti of the University of Milan, Italy, when he was assigned to the Ames Research Center as a resident Research Associate sponsored by the National Academy of Sciences.[2] The experiment was designed to study the adaptability of the otolith to sustained weightlessness, to provide information for manned space flight. Originally planned in 1966 to be included on an early Apollo mission, the experiment was deferred when that mission was canceled. In late 1967 authorization was given to orbit the FOE when a supporting spacecraft could be designed.[3] The project, part of NASA's Human Factor Systems program, was officially designated "OFO" in 1968.[4] After a series of delays, *OFO* was orbited 9 November 1970.

OGO. An acronym for "Orbiting Geophysical Observatory," the name was derived from NASA's concept for an observatory-class satellite. In late 1959 and early 1960, the concept evolved from that of a larger general-purpose scientific satellite (as opposed to the special-purpose Explorers), which would be a standardized spacecraft housing a variety of instruments to be flown regularly on standardized trajectories. "Orbiting Observatory" became the term used for this class of spacecraft, and "Orbiting Geophysical Observatory" developed as a functional description for this particular satellite.

The names "EGO" and "POGO" also were developed during this period to apply to OGO satellites in particular orbital trajectories: highly eccentric (Eccentric Geophysical Observatory) and polar orbit (Polar Orbiting Geophysical Observatory).[1] Between 1964 and 1969 NASA orbited six OGO satellites and results from the successful OGO program included the first global survey by satellite of the earth's magnetic field.

OSO. An acronym for "Orbiting Solar Observatory," OSO evolved from the NASA concept for larger, general-purpose spacecraft for scientific ex-

Artist's concept of an OGO spacecraft in orbit.

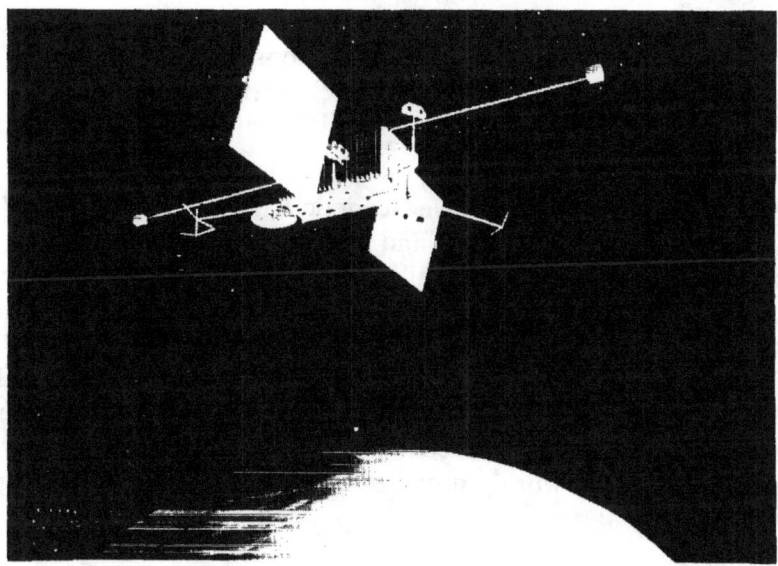

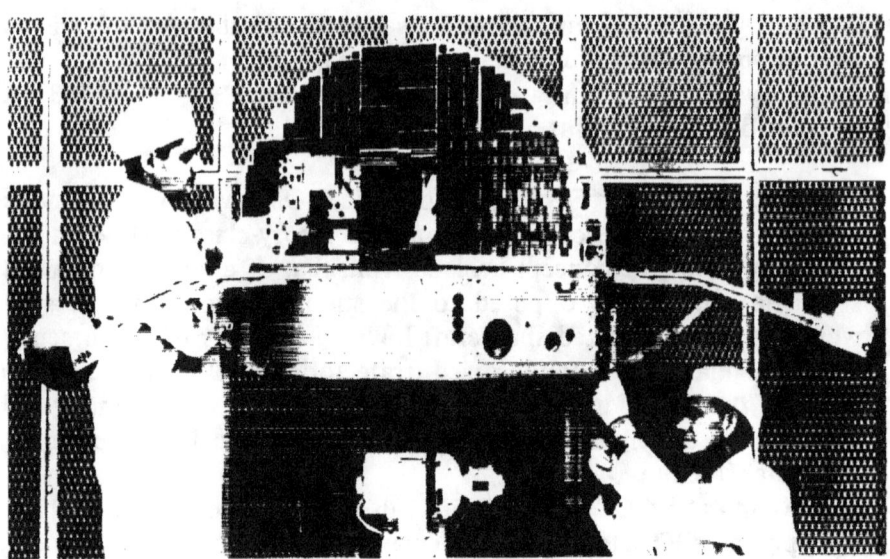

OSO 3 *undergoing spin-balance tests before launch.*

periments (see OGO). The name was a functional description of the satellite, indicating it was of the orbiting-observatory class of satellites whose purpose was to measure phenomena of the sun.[1] *OSO 1*, launched 7 March 1962, was the first satellite in the "Orbiting Observatory" series to be placed in orbit. *OSO 7* was launched 29 September 1971.

The OSO satellites were designed to provide observations of the sun during most of its 11-year cycle. Results included the first full-disc photograph of the solar corona, the first x-ray observations from a spacecraft of a beginning solar flare and of solar "streamers"—structures in the corona—and the first observations of the corona in white light and extreme ultraviolet.

PAGEOS. The acronym for "Passive Geodetic Earth Orbiting Satellite"[1] came into use among project officials and found its way into documents through common use.[2] "PAGEOS" paralleled the name "GEOS" that designated the active (instrumented) geodetic satellites in the Explorer series. In August 1964 NASA approved Langley Research Center's proposal for the PAGEOS project. *Pageos 1*, a balloon 30 meters in diameter—similar to the Echo balloon satellite—achieved orbit and inflated 23 June 1966. The uninstrumented (passive) satellite reflected sunlight and, photographed by ground stations around the world, provided a means of precision mapping the earth's surface.[3]

PAGEOS 1 *inflation test, above, and a Pegasus satellite with wings outspread in an artist's drawing.*

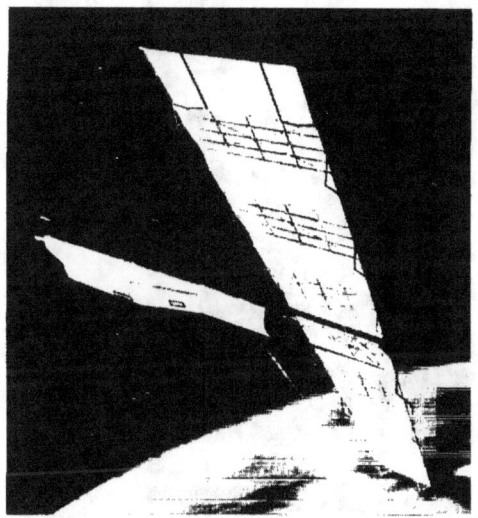

PEGASUS. The outstanding feature of the Pegasus satellites was their huge winglike panels, 96 meters tip to tip, sweeping through space to determine the rate of meteoroid penetrations. The program office said when choosing from proposed names that the spacecraft, to be the heaviest yet orbited, would be "somewhat of a 'horse' as far as payloads are concerned" and there could be "only one name for a horse with wings"—Pegasus, the name of the winged flying horse of ancient Greek mythology.[1]

The original suggestion for the name had come from an employee of the spacecraft contractor, Fairchild Stratos Corporation. The contractor, with the concurrence of the NASA Office of Space Vehicle Research and Technology and Marshall Space Flight Center, had held an in-house competition in 1963 to select a name for the project. From more than 100 suggestions by Fairchild Stratos employees, the NASA program office recommended the name "Pegasus" to the Project Designation Committee. The Committee approved the selection in July 1964* and NASA announced the name in August.[2]

A Relay satellite, below; and San Marco 1, *at right, in checkout at Wallops.*

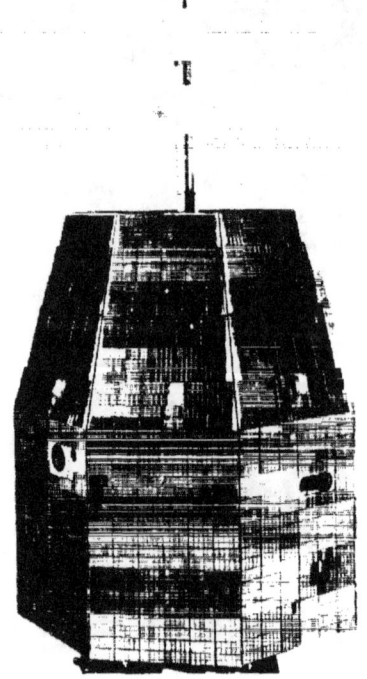

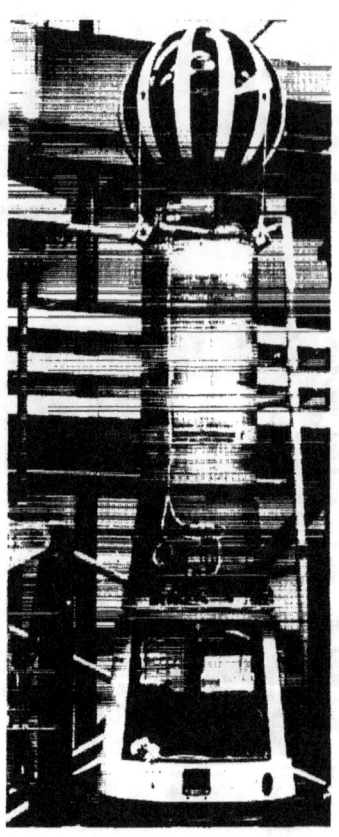

---

*The NASA Project Designation Committee originally agreed on "Project Pegasus" as the name for the experiments before launch. The satellites were to be supplanted with an Explorer designation in orbit. (Julian W. Scheer, Assistant Administrator for Public Affairs, NASA, memorandum for Raymond L. Bisplinghoff, Associate Administrator for Advanced Research and Technology, with concurrence of Robert C. Seamans, Jr., Associate Administrator, 6 July 1964.)

SATELLITES

Three Pegasus satellites were placed in orbit, all by Saturn I launch vehicles: *Pegasus 1* on 16 February 1965, *Pegasus 2* on 25 May 1965, and *Pegasus 3* on 30 July 1965.

RELAY. NASA's medium-altitude, active-repeater communications satellite was formally named "Relay" in January 1961* at the suggestion of Abe Silverstein, NASA's Director of Space Flight Programs.[1]. The name was considered appropriate because it literally described the function of an active-repeater comsat: the satellite received a signal, amplified it within the satellite, and then relayed the signal back toward the earth.[2] *Relay 1,* orbited 13 December 1962, and its successor *Relay 2,* orbited 21 January 1964, both demonstrated the feasibility of this kind of communications satellite. After its research role was completed, *Relay 2* was turned over to the Department of Defense to assist in military communications over the Pacific.

SAN MARCO. The Italian space program was conceived in 1960 by Professor Luigi Broglio, Professor Carlo Buongiorono, and Dr. Franco Fiorio.[1] By 1962 they and their colleagues had decided that an ocean platform in nonterritorial waters should serve as the base for launching their satellite booster. ENI, Italy's state-owned oil industry, made available a suitable platform, which happened to be named "San Marco" (Saint Mark). The name "San Marco" grew into the designation for the entire cooperative space project—including preparatory phases not associated directly with the sea-based launch site. Professor Broglio was particularly pleased to adopt the name for the project because Saint Mark was the patron saint of Venice, his birthplace. Saint Mark was also the patron of all who sailed the sea.[2]

The San Marco project was a cooperative effort of NASA and the Italian Space Commission, with NASA providing launch vehicles, use of its facilities, and training of Italian personnel. On 15 December 1964, the San Marco Scout 1 booster, launched from Wallops Station by an Italian crew, orbited the Italian-designed-and-built *San Marco 1* satellite. The launch was the first satellite launch in NASA's international cooperation program that was conducted by non-U.S. personnel and was the first Western European satellite launch. *San Marco 2* was launched into equatorial orbit 26 April 1967 from

---

*Eight months earlier, Lloyd E. Jones, Jr., a member of the NASA Ad Hoc Committee to Name Space Projects and Objects, had suggested to the Committee a group of names for applications satellites. The Committee approved the name "Relay" for an active comsat 19 May 1960. (NASA Ad Hoc Committee to Name Space Projects and Objects, minutes of meeting 19 May 1960.)

the San Marco platform in the Indian Ocean. *San Marco 3*, launched 24 April 1971, was the third satellite orbited from the platform (the second had been NASA's *Explorer 42*, launched 12 December 1970). *San Marco 4* was launched from the platform 18 February 1974.

The San Marco satellites were scientific satellites designed to conduct air-density experiments using a variety of instruments; in addition, *San Marco 1* and *2* measured ionospheric characteristics related to long-range radio transmission.

SEASAT. The name of the "sea satellite"—"Specialized Experimental Applications Satellite," shortened to the acronym "SEASAT"—was chosen before the program was officially established. A 1969 conference of scientists and representatives from the National Oceanic and Atmospheric Administration, Department of Defense, NASA, other Government agencies, universities, and scientific institutions met at Williams College, Williamstown, Massachusetts, to review activities needed in the earth and ocean physics fields. The conference identified a number of activities, including satellite projects. SEASAT and LAGEOS (see LAGEOS) were among them, the names growing out of the thinking of a number of the participants and fitting the tasks of the satellites within NASA's earth and ocean physics applications program (EOPAP). [1]

After studies and definition of requirements in cooperation with numerous Government agencies and private institutions, through the SEASAT User Working Group, NASA introduced SEASAT as a "new start" in its Fiscal Year 1975 program. The new satellite was scheduled for 1978 launch, following technological evolution of equipment on the Skylab and GEOS–C missions; it would be the first devoted entirely to studying the oceans. SEASAT was to circle the globe 14½ times a day to observe weather and sea conditions of all the earth's oceans with accurate micro-wave devices. Information was to be distributed to a large user community for predicting weather, routing shipping, and issuing disaster warnings. This first satellite in the program was to be a proof-of-concept spacecraft for later operational missions.

SIRIO. In March 1970 NASA and the Italian National Research Commission signed a memorandum of understanding providing for the reimbursable NASA launch of Italian scientific spacecraft. [1] The first satellite planned for launch under this agreement was SIRIO—an acronym for "Satellite Italiano Ricerche Orientate (Italian Research-Oriented Satellite)." [2] Tentatively

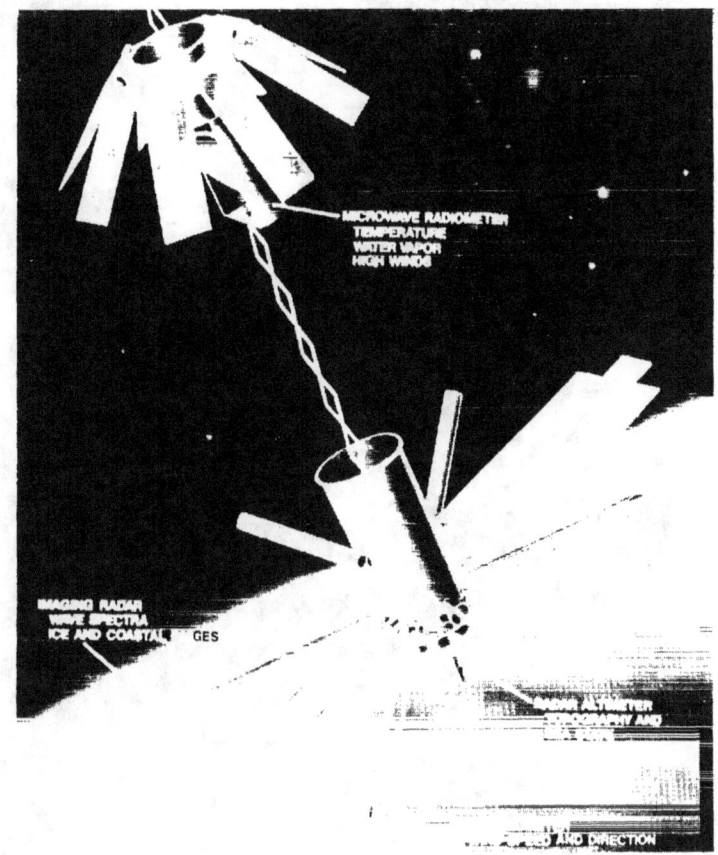

An artist's concept of SEASAT, a satellite to study oceans.

scheduled for launch in 1975, SIRIO would conduct telecommunications, technology, and scientific experiments from synchronous orbit.

SMS. An operational satellite system that could provide continuous observation of weather conditions from a fixed position above the earth had been under study since the first weather satellites were launched in the early 1960s. Studies of the requirements for a stationary weather satellite were begun in early 1960 and the proposed project was named for Aeros, ancient Greek god of the air.[1] Conceived as the third phase of a program consisting of Tiros and planned Nimbus satellites, Aeros would be a synchronous satellite in equatorial orbit that could track major storms as well as relay cloud-cover photographs of a large portion of the earth.[2]

69

The second SMS satellite, erected atop its Delta launch vehicle at Kennedy Space Center's Launch Complex 17.

By late 1962 the name Aeros had been dropped in favor of the more functional designation "SMS," an acronym for "Synchronous Meteorological Satellite."[3] Meanwhile, studies were being made of a Tiros spacecraft ("Tiros-K") that could be modified for a near-synchronous orbit to determine the capability of an SMS.[4] Tiros-K was subsequently canceled in 1965 as development plans for the ATS satellites permitted the inclusion of experiments to test the proposed instrumentation for the SMS.[5]

After the successful photographic results of *ATS 1* and *3*, two experimental SMS satellites were approved and tentatively planned for launch. SMS–A and SMS–B, funded by NASA, would be prototypes for the later operational satellites funded by the National Oceanic and Atmospheric Administration (NOAA). Following launch and checkout by NASA, both satellites were to be turned over to NOAA for use in the National Operational Meteorological Satellite System (NOMSS).* Successive satellites in the series would be designated "GOES"—an acronym for "Geostationary Operational Environmental Satellite"—by NOAA.[6] An operational system of two or more SMS satellites and a single ITOS spacecraft could provide the coverage required for accurate long-range weather forecasts.

SMS–A became *SMS 1* on launch into orbit 17 May 1974 and supported the international Global Atmospheric Research Program's Atlantic Tropical Experiment (GATE) before becoming part of NOAA's operational system late in the year. SMS–B and GOES–A (SMS–C) were scheduled for 1975 launch. The European Space Research Organization, Japan, and the U.S.S.R. were planning to launch their own geostationary satellites during the decade to complement the SMS system for global use.[7]

SPHINX. Planned as one of NASA's smallest scientific satellites, the 113-kilogram SPHINX took its name from the acronym for "Space Plasma High Voltage Interaction Experiment." It was to be launched pickaback on the proof flight of the newly combined Titan III–Centaur launch vehicle, along with a dynamic model of the Viking spacecraft. The planned year-long mission was to measure effects of charged particles in space on high-voltage solar cells, insulators, and conductors. Data would help determine if future spacecraft could use high-voltage solar cells, instead of the present low-voltage cells, to operate at higher power levels without added weight or cost. The Centaur stage failed on launch 11 February 1974, however, and the satellite was destroyed.[1]

---

*NOMSS was also known by NOAA as the National Operational Environmental Satellite System.

Flight model of the French-German Symphonie *above. A* Syncom *fires its apogee motor in orbit in the photographic composite.*

SYMPHONIE. The experimental Franco-German Symphonie communications satellites were designed and built in Europe for launch by NASA with launch vehicles and services paid for by France and West Germany.

Two satellites were developed by the joint Consortium Industriel France-Allemand pour le Satellite Symphonie (CIFAS) under the direction of the French space agency Centre National d'Études Spatiales (CNES) and the West German space agency Gesellschaft für Weltraumforschung (GfW). The three-axis stabilized satellites were to test equipment for television, radio, telephone, telegraph, and data transmission from synchronous orbit, 35 900 kilometers above the equator. They were planned for launch from

French Guiana on the Europa II launch vehicle, but when the European Launcher Development Organization (ELDO) canceled its vehicle project the countries turned to NASA. The contract for NASA launch services was signed in June 1974.[1]

In 1967 France had a stationary (synchronous) orbit communications satellite, SAROS (Satellite de Radiodiffusion pour Orbit Stationnaire), in the design stage and West Germany was about to begin designing its Olympia satellite. The two nations agreed in June 1967 to combine their programs in a new joint effort. Participants in the 1967 discussions in Bonn—the Federal Republic of Germany's capital on the Rhine River—sought a new name for the joint satellite just before the agreement was signed. Gerard Dieulot, technicaldirector of the French program, was reminded of the German composer Robert Schumann by the name of French Minister Maurice Schumann, negotiator for France. The new accord in the Rhine Valley, Dieulot suggested, was a "symphony by Schumann." "Symphonie," the French spelling of the word coming originally from the Latin and Greek "symphonia," "harmony" or "agreement," was adopted when the Franco-German satellite agreement was signed in June.[2]

NASA launched *Symphonie 1* (Symphonie-A before launch) into orbit from Eastern Test Range 18 December 1974. Symphonie-B was scheduled for September 1975 launch.

SYNCOM. A word coined from the first syllables of the words "synchronous communications," "Syncom" referred to communications satellites in synchronous earth orbit. The name was devised by Alton E. Jones of NASA Goddard Space Flight Center. Early in August 1961, when he was working on the preliminary project development plan, he decided that a name was required before the plan could go to press the next day. He invented the name "Syncom." [1] Before the end of August, NASA Headquarters officials had approved the preliminary plan and NASA had issued a press release using the name.[2]

Three Syncom satellites were developed and launched. After a launching success but communications failure with *Syncom 1* (14 February 1963), *Syncom 2* was launched 26 July 1963, into the first synchronous orbit and *Syncom 3,* launched 19 August 1964, was put into the first truly stationary orbit. The Department of Defense participated in Syncom research and development, providing ground stations and conducting communications experiments. Early in 1965, after completing the research and development program, NASA transferred use of the two Syncom satellites to the Department of Defense.

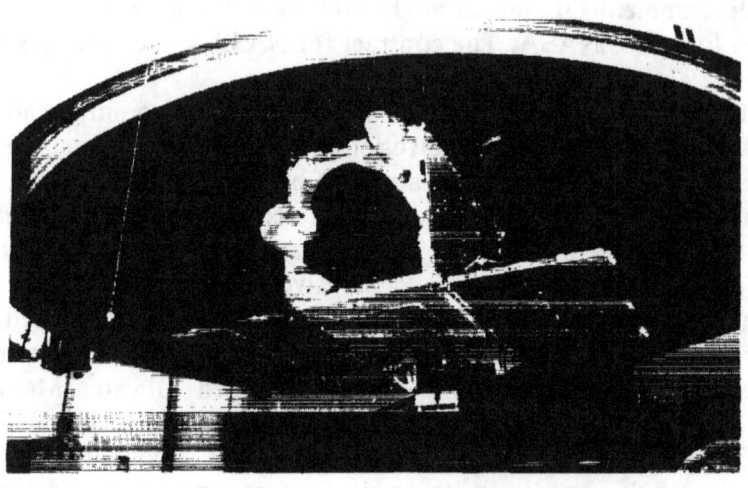

Structural model of TD–1A *(above) showing the telescope, suspended for a space simulation test (ESC photo). At left, Telesat satellite* Anik 1 *in production (Hughes Aircraft Co. photo); below a composite photograph of the* Telstar 1 *spacecraft in space.*

TD. "TD," an abbreviation for the U.S. Thor-Delta launch vehicle, was the name given to a satellite project by the European Space Research Organization (ESRO).[1] *TD-1A,* a solar astronomy satellite designed to carry a variety of instruments including a large telescope, was launched by NASA 11 March 1972. Under a 1966 memorandum of understanding with ESRO, NASA was reimbursed for the launch.[2]

Proposals for the satellite, then unnamed, had been discussed at an astronomy colloquium soon after the formal establishment of ESRO in March 1964.[3] By 1965, ESRO had planned a series of TD satellites and in 1967, after several program delays, signed a contract with NASA for the launch of two satellites, TD-1 and TD-2.[4] In April 1968, however, ESRO announced the cancellation of both satellites because of problems in financing. The project was later reinstated and a second contract for a single Thor-Delta launch was signed with NASA in June 1970. The satellite was subsequently redesignated "TD-1A" because it differed from the two earlier configurations and combined the TD-2 design with several experiments originally planned for TD-1.[5]

TELESAT. In early 1969 the Canadian Ministry of Communications proposed plans for a satellite system that could be used entirely for domestic communications.[1] The system would be managed and operated by Telesat ("Telecommunications Satellite") Canada, a new corporation supported by industry, government, and public investment. The first two satellites in the system, Telesat-A and B, would be launched into synchronous equatorial orbit and be capable of relaying TV, telephone, and data transmissions throughout Canada. Under an agreement with Telesat, NASA would provide the Thor-Delta launch vehicles and be reimbursed for the satellite launches.[2]

In orbit each Telesat satellite would be designated "Anik," the Eskimo word for "brother."[3] *Anik 1* was launched into orbit 9 November 1972 and *Anik 2* on 20 April 1973. Anik 3 was scheduled for 1975.

TELSTAR. A contraction of "telecommunications" and "star," the name "Telstar" designated the active communications satellites developed by American Telephone & Telegraph Company. In November 1961, at the request of AT&T's Bell Telephone Laboratories, NASA endorsed the selection of "Telstar" as a name for the project.[1] NASA was responsible for launching, tracking, and data acquisition for the AT&T-built satellites on a cost-reimbursable basis. *Telstar 1,* the first active-repeater communications satellite, was the first privately funded satellite and relayed the first live transatlantic telecast after 10 July 1962 launch. It was followed by the equally successful *Telstar 2,* 7 May 1963.

ITOS–B being checked out before launch.

TIROS, TOS, ITOS. The Tiros meteorological satellite, which provided weather data from high above the earth's cloud cover, was given a name that described its function. In mid-1958, the Department of Defense's Advanced Research Projects Agency (initiator of the project) requested the Radio Corporation of America (contractor for the project) to supply a name for the satellite. RCA personnel concocted the name "TIROS," an acronym derived from the descriptive title "Television and Infra-Red Observation Satellite." [1] The name eventually came to be written "Tiros" as it was used in other acronyms.

SATELLITES

In April 1959 responsibility for the Tiros research and development program was transferred from the Department of Defense to NASA, and on 1 April 1960 *Tiros 1* was launched into orbit. Meteorologists were to receive valuable data—including more than 5 000 000 usable cloud pictures—from 10 Tiros weather satellites.[2] By early 1964 NASA had orbited *Tiros 1* through *Tiros 8* and the U.S. Weather Bureau was making operational use of the meteorological data from them. These satellites were able to photograph about 20 percent of the earth each day.

On 28 May 1964 NASA and the Weather Bureau announced a plan for an operational meteorological satellite system based on Tiros research and development. They called the system "TOS"—an acronym for "Tiros Operational Satellite." In accordance with the NASA–USWB agreement, *Tiros 9* was a NASA-financed, modified Tiros satellite, orbited to test the new "cartwheel" configuration on which the TOS would be based. *Tiros 10* was a USWB-financed, Tiros satellite similar to *Tiros 9,* orbited to continue testing the TOS concept. Early in 1966 NASA orbited the two operational satellites in the TOS system—financed, managed, and operated by the Weather Bureau, by then an agency of the new Environmental Science Services Administration (ESSA). Upon their successful orbit, ESSA designated the TOS satellites *ESSA 1* and *ESSA 2*—ESSA in this case being an acronym for "Environmental Survey Satellite." [3] These two satellites provided continuous cloud-cover pictures of the entire sunlit portion of the earth at least once daily.

In 1966 NASA announced plans for a design study of an improved TOS spacecraft that would be twice as large as the previous TOS satellites. This spacecraft would be able to scan the earth's nighttime cloud cover and would more than double the daily weather coverage obtained in the TOS series of ESSA satellites.[4] The first satellite in the Improved Tiros Operational Satellite (ITOS) series, *ITOS 1*—launched 23 January 1970—was a joint project of NASA and ESSA. With the exception of *ITOS 1,* spacecraft in the ITOS series would be funded by ESSA.[5]

On 3 October 1970, ESSA was combined with the major Federal programs concerned with the environments of the sea and air; programs from four departments and one agency were consolidated to form the National Oceanic and Atmospheric Administration (NOAA) in the Department of Commerce. The first operational ITOS spacecraft funded by NOAA—designated *NOAA 1* in orbit—was launched 11 December 1970. *NOAA 4* (ITOS–G) was put into orbit 15 December 1974 to join the still orbiting *NOAA 2* and *3* (launched 15 October 1972 and 6 November 1973) in obtaining global cloud-cover data day and night and global measurements of the earth's atmospheric structure for weather prediction.

VANGUARD. The name "Vanguard" applied to both the first satellite series undertaken by the United States and to the launch vehicle developed to orbit the satellites. In the spring of 1955, scientific interest in orbiting an artificial earth satellite for International Geophysical Year (1 July 1957 to 31 December 1958) was growing. Several launch vehicle proposals were developed for placing a U.S. satellite in orbit. The proposal chosen in August

Vanguard 2 *atop its satellite launch vehicle.*

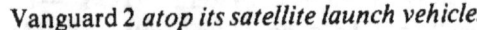

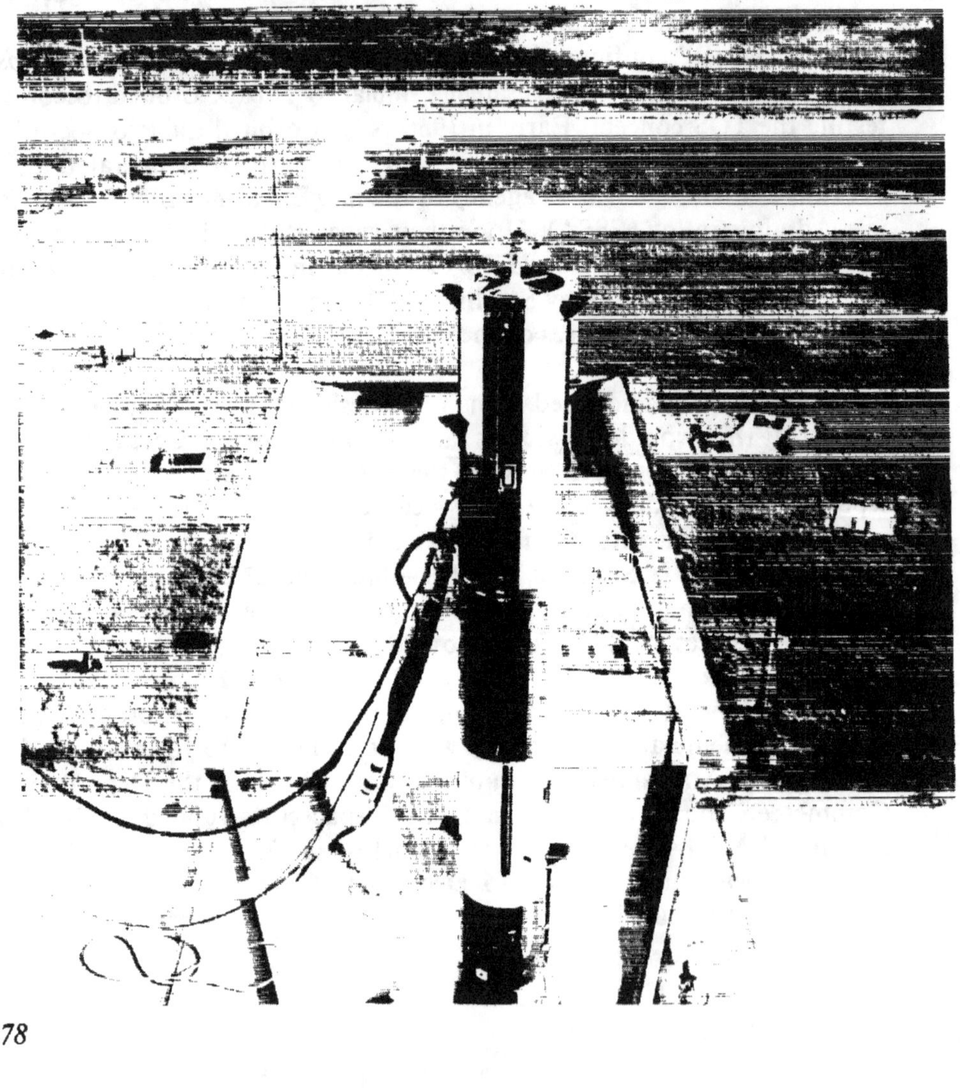

1955 to be the U.S. satellite project for the IGY was the one offered by the Naval Research Laboratory (NRL), based on Milton W. Rosen's concept of a new launch vehicle combining the Viking first stage, Aerobee second stage, and a new third stage.[1] Rosen became technical director of the new project at NRL.

The name "Vanguard" was suggested by Rosen's wife, Josephine. Rosen forwarded the name to his NRL superiors, who approved it. The Chief of Naval Research approved the name 16 September 1956.[2] The word denoted that which is "out ahead, in the forefront."

Vanguard 1, a 1.5-kilogram scientific satellite, was orbited 17 March 1958, although ironically it was not the first U.S. satellite (*Explorer 1* had been launched into orbit by the Army 31 January 1968). The NRL project Vanguard team was transferred to NASA when the space agency was established 1 October 1958. *Vanguard 1* was followed in 1959 by *Vanguard 2* and *3.* Scientific results from this series included the first geodetic studies indicating the earth's slightly "pear" shape, a survey of the earth's magnetic field, the location of the lower edge of the earth's radiation belts, and a count of micrometeorite impacts.

WESTAR. Westar satellites were commercial communications satellites owned and operated by Western Union Telegraph Company and launched by NASA under a contract, to form the first United States domestic communications satellite system.

As early as 1966, Western Union petitioned the Federal Communications Commission for permission to build a domestic satellite system to relay telegraph traffic. The FCC was then making a detailed study of the need for such a system in response to requests from several organizations. When the FCC decided in 1970 to invite applications, Western Union was the first to respond, proposing a high-capacity multipurpose system to serve all 50 states.[1] The company won approval in January 1973 to build the first U.S. system, with authorization for three satellites. Hughes Aircraft Company was to build the comsats (or "domsats," as the press began to call them) and NASA signed a contract with Western Union in June 1973, agreeing to provide launch services, with reimbursement for the Thor-Delta launch vehicles and costs.[2]

Western Union asked its employees to suggest a name for the new satellites. From the suggestions, "Westar" was chosen—combining part of the company's name with "star," a reference to a body in space, or satellite.[3]

Westar 1 ("Westar-A" before launch) was orbited 13 April 1974 and began commercial operation 16 July. As a new postal service, *Westar 1* relayed the first satellite "Mailgrams" in 1974, from New York to Los

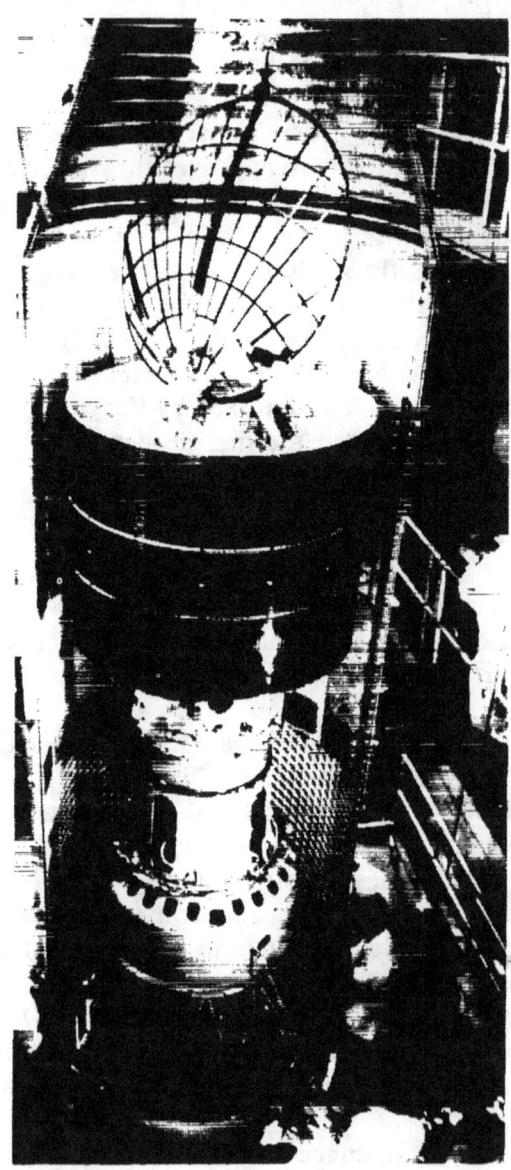

Westar 1 *being encapsulated in its payload shroud on top the Delta launch vehicle at Kennedy Space Center's Launch Complex 17.*

Angeles at the speed of light. *Westar 2* was launched 10 October 1974 and Westar-C was held as a spare. In synchronous orbit, each drum-shaped satellite could relay 12 color TV channels, up to 14 400 one-way telephone circuits, or multiple data channels.

III

SPACE PROBES

Whirlpool-shaped galaxy.

SPACE PROBES

Unmanned instrumented probes obtain scientific information about the moon, other planets, and the space environment. Probes are differentiated from sounding rockets in that they attain at least 6400-kilometer altitudes. When a probe is launched on an escape trajectory—attaining sufficient velocity to travel beyond the earth's gravitational field—it becomes, in effect, a satellite of the sun. The Lunar Orbiter probes, however, were sent into orbit around the earth's natural satellite, the moon.

First serious consideration of the concept of a space probe can be attributed to Dr. Robert H. Goddard, American rocket pioneer. As early as 1916, Goddard's calculations of his theoretical rocket and his experiments with flash powders led him to conclude that a rocket-borne payload exploding on the moon could be detected from earth.[1] On 20 September 1952 a paper entitled "The Martian Probe," presented by E. Burgess and C. A. Cross to the British Interplanetary Society, gave the term "probe" to the language.[2]

In May 1960—at the suggestion of Edgar M. Cortright, Assistant Director of Lunar and Planetary Programs—NASA adopted a system of naming its space probes. Names of lunar probes were patterned after land exploration activities (the name "Pioneer," designating the early series of lunar and related space probes, was already in use). The names of planetary mission probes were patterned after nautical terms, to convey "the impression of travel to great distances and remote lands." Isolated missions to investigate the space environment were "assigned the name of the mission group of which they are most nearly a part."[3] This 1960 decision was the basis for naming Mariner, Ranger, Surveyor, and Viking probes.

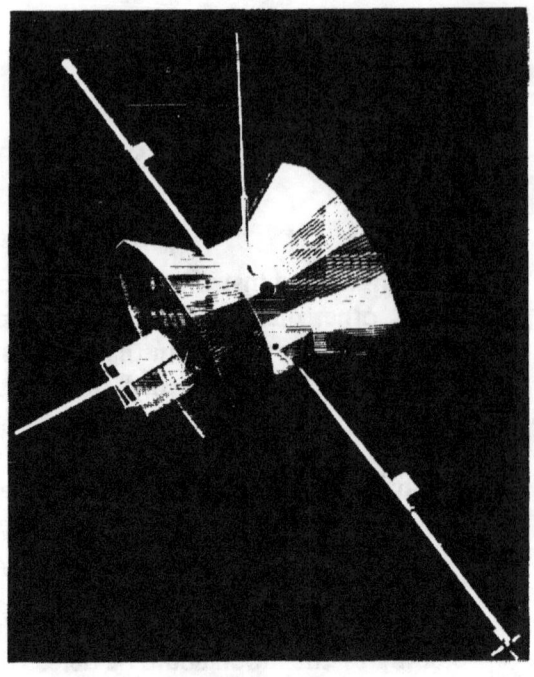

U.S.-German Helios 1 *solar probe on the spin table for system test and checkout before mating to the Titan III launch vehicle.*

HELIOS. In June 1969 NASA and the German Ministry for Scientific Research (BMwF) agreed to a joint project for launching two probes, in 1974 and 1975, to study the interplanetary medium and explore the near-solar region. The probes would carry instruments closer to the sun than any previous spacecraft, approaching to within 45 million kilometers.[1]

The project was designated "Helios," the name of the ancient Greek god of the sun, by German Minister Karl Kaesmeier. The name had been suggested in a telephone conversation between Minister Kaesmeier and Goddard Space Flight Center's Project Manager, Gilbert W. Ousley, in August 1968.[2] NASA had previously used the name for the Advanced Orbiting Solar Observatory (AOSO), canceled in 1965, which was to have performed similar experiments.[3] The Helios probes were to be launched on Titan III–Centaur vehicles.

NASA launched West German-built *Helios 1* into orbit of the sun 10 December 1974. Helios-B was scheduled for 1976 launch.

LUNAR ORBITER. The name "Lunar Orbiter" was a literal description of the mission assigned to each probe in that project: to attain lunar orbit, whence it would acquire photographic and scientific data about the moon. Lunar Orbiter supplemented the Ranger and Surveyor probe projects, providing lunar data in preparation for the Apollo manned landings and the Surveyor spacecraft softlandings. [1]

The name evolved informally through general use. NASA had had under consideration plans for a Surveyor spacecraft to be placed in orbit around the moon. This Surveyor was called "Surveyor Orbiter" to distinguish it from those in the lunar-landing series. When the decision was made to build a separate spacecraft rather than use Surveyor, the new probe was referred to simply as "Orbiter" or "Lunar Orbiter." [2]

Five Lunar Orbiter flights launched in 1966 and 1967 made more than 6000 orbits of the moon and photographed more than 99% of the lunar surface, providing scientific data and information for selecting the Apollo manned landing sites. Tracking data increased knowledge of the moon's gravitational field and revealed the presence of the lunar mascons. [3]

Scale models of a Lunar Orbiter spacecraft and the moon in the top photo demonstrate the approach to within 48 kilometers of the lunar surface. Below, a portion of the first closeup of the lunar crater Copernicus, taken 23 November 1966 by Lunar Orbiter 2.

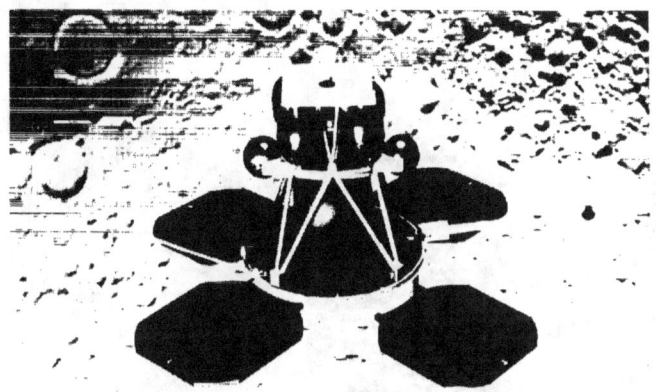

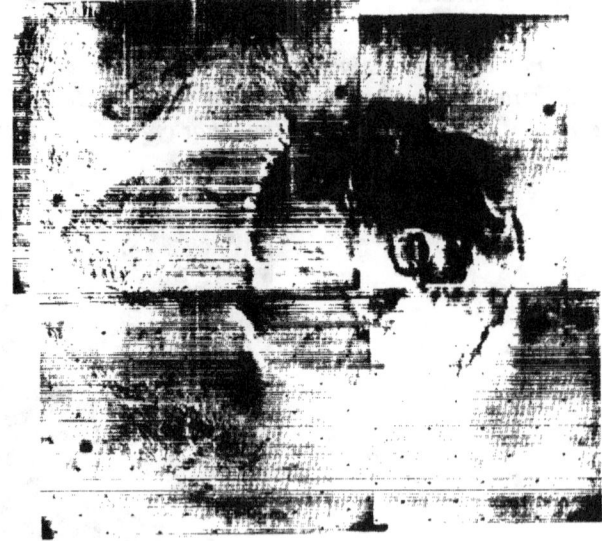

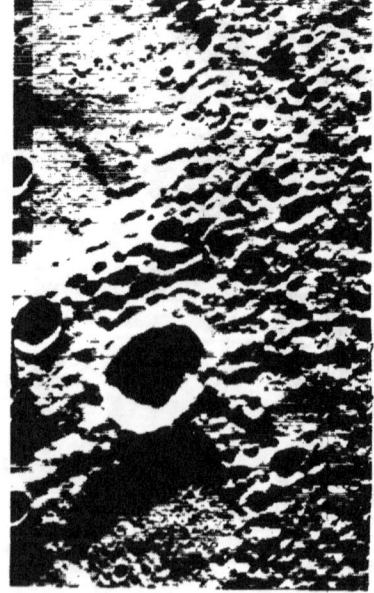

Mariner 9 *spacecraft with thermal blanket covering the retro engine at top. Nix Olympica, gigantic volcanic mountain on Mars, photographed by* Mariner 9 *in January 1972, above.* Mariner 10 *photographed the densely cratered surface of Mercury on 29 March 1974, at 18 200 kilometers from the planet.*

SPACE PROBES

MARINER. The space probes to investigate the vicinities of the earth's planetary neighbors, Venus and Mars, and eventually Mercury, Jupiter, and Saturn, were designated the "Mariner" series. The name was adopted in May 1960 as part of the Cortright system of naming planetary missions from nautical terms.[1]

Mariner spacecraft made a number of record-setting missions, from the early years of the project. On 14 December 1962 NASA's *Mariner 2* came within 34 900 kilometers of Venus, climaxing a four-month space flight that provided new scientific data on interplanetary space and Venus. On 14 July 1965, after seven months of interplanetary flight, *Mariner 4* took the first close look at Mars from outside the earth's atmosphere, returning high-quality photographs and scientific data.

On 19 October 1967 *Mariner 5* flew within 4000 kilometers of Venus, obtaining additional information on the nature and origin of the planet and on the interplanetary environment during a period of increased solar activity. During 1969, *Mariner 6* and *7* continued the investigation of the Martian atmosphere, flying within 3500 kilometers of the planet. Following the unsuccessful Mariner 8 launch attempt,[*] *Mariner 9* was launched 30 May 1971 and put into orbit around Mars on 13 November 1971—the first man-made object to orbit another planet. *Mariner 9* photographed the moons of Mars, mapped 100 percent of the planet, and returned data proving it was geologically and meteorologically alive.

Mariner 10, launched 3 November 1973, flew past Venus in February 1974 to a March 1974 encounter with Mercury, for the first exploration of that planet. The spacecraft's trajectory around the sun swung it back for a second encounter with Mercury in September 1974 and would return it for a third in March 1975. Venus data gave clues to the planet's weather system, suggested the planet's origin differed from the earth's, and confirmed the presence of hydrogen in its atmosphere. Mercury data revealed a strong magnetic field, a tenuous atmosphere rich in helium, a cratered crust, and possibly an iron-rich core; it brought new insight into the formation of the terrestrial planets.

Two Mariner Jupiter-Saturn probes were planned for launch in 1977 to study the environment, atmosphere, and characteristics of those planets.[2]

---

[*]Mariner H was designated Mariner 8 by NASA Associate Administrator John E. Naugle because of pressure from the press for easier identification. This designation was a departure from past precedent of assigning a number to spacecraft only after a successful launch. (NASA, Mariner Mars 1971 Project Office, telephone interview, 4 June 1971.)

Pioneer 11 *spacecraft during checkout with a mockup of the launch vehicle's third stage, before launch. Jupiter's red spot and a shadow of the moon Io with the planet's cloud structure were photographed by* Pioneer 10 *on 1 December 1973.*

PIONEER. "Pioneer" was chosen as the name for the first U.S. space probe, *Pioneer 1,* launched 11 October 1958, as well as for the following series of lunar and deep space probes. The Pioneer series had been initiated for the International Geophysical Year by the Department of Defense's Advanced Research Projects Agency (ARPA), which assigned execution variously to the Air Force Ballistic Missile Division (AFBMD) and to the Army Ballistic Missile Agency (ABMA). Upon its formation in October 1958, NASA inherited responsibility for—and the name of—the probes.[1]

Credit for naming the first probe has been attributed to Stephen A. Saliga, who had been assigned to the Air Force Orientation Group, Wright-Patterson AFB, as chief designer of Air Force exhibits. While he was at a briefing, the spacecraft was described to him as a "lunar-orbiting vehicle with an infrared scanning device." Saliga thought the title too long and lacked theme for an exhibit design. He suggested "Pioneer" as the name of

the probe since "the Army had already launched and orbited the Explorer satellite and their Public Information Office was identifying the Army as 'Pioneers in Space,'" and by adopting the name the Air Force would "make a 'quantum jump' as to *who* really [were] the 'Pioneers in Space.'" [2]

The first series of Pioneer spacecraft was flown between 1958 and 1960. *Pioneer 1, 2,* and *5* were developed by Space Technology Laboratories, Inc., and were launched for NASA by AFBMD. *Pioneer 3* and *4* were developed by the Jet Propulsion Laboratory and launched for NASA by ABMA. In 1960 *Pioneer 5* transmitted the first solar flare data and established a communications distance record of 36.2 million kilometers.

With the launch of *Pioneer 6* (Pioneer A in the new series) in December 1965, NASA resumed the probes to complement interplanetary data acquired by Mariner probes. *Pioneer 7, 8,* and *9,* second-generation spacecraft launched between 1966 and 1968, continued the investigation of the inter-planetary medium.

Between 1965 and 1967 NASA had been studying the concept for a space probe known as the "Galactic Jupiter Probe," or "Advanced Planetary Probe," that would investigate solar, interplanetary, and galactic phenomena in the outer region of the solar system.[3] By 1968 NASA had included the probe in the Pioneer series, designating two such probes Pioneer F and G.[4]

Pioneer 10 (Pioneer F), launched in March 1972, became the first spacecraft to cross the Asteroid Belt. It flew by Jupiter in December 1973, returning more than 300 closeup photos of the planet and its inner moons as well as data on its complex magnetic field and its atmosphere. Accelerated by Jupiter's gravity, the probe was to reach the orbit of Saturn in 1976 and the orbit of Uranus in 1979; it was expected to become in 1987 the first spacecraft to escape the solar system.

Pioneer 11 (Pioneer G), launched in April 1973, crossed the Asteroid Belt, skimmed by Jupiter three times closer to the planet than *Pioneer 10* had, and was thrown by Jupiter's gravity toward Saturn. The spacecraft sent back the first photos of Jupiter's poles and information on the atmosphere, the equator regions, and the moon Callisto. On the night of 2 December 1974, when *Pioneer 11* set its new course for Saturn, NASA renamed the probe *Pioneer Saturn.*[5] It was to pass close by Saturn in the fall of 1979.

Two Pioneer Venus spacecraft, an orbiter and a multiprobe lander, were to gather detailed information on the atmosphere and clouds of Venus in 1978. The lander was to release four probes to the planet's surface.[6]

RANGER. A probe series to gather data about the moon, Ranger was assigned its name in May 1960 because of the parallel to "land exploration activities."[1] NASA had initiated Project Ranger—then unnamed—in December 1959, when it requested Jet Propulsion Laboratory (JPL) to study spacecraft design and a mission to "acquire and transmit a number of images of the lunar surface."[2] In February 1960 Dr. William H. Pickering, JPL Director, recommended that NASA Headquarters approve the name "Ranger" used by JPL for the project. The name had been introduced by the JPL program director, Clifford D. Cummings, who had noticed while on a camping trip that his pick-up truck was called "Ranger." Cummings liked the name and, because it referred to "land exploration activities," suggested it as a name for the lunar impact probe. By May 1960 it was in common use.[3]

Ranger 7 *before 28 July 1964 launch to the moon, at left. The television picture of craters on the lunar surface was taken by* Ranger 9 *before impact 24 March 1965.*

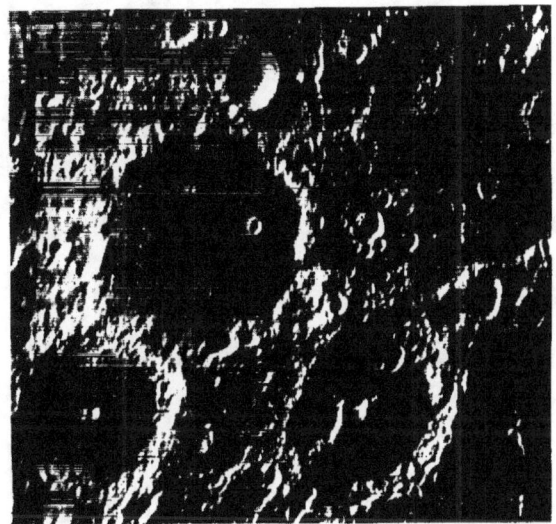

The first U.S. spacecraft to hit the moon was *Ranger 4,* launched 23 April 1962. *Ranger 7, 8,* and *9,* flown 1964–1965, provided thousands of close-up photographs of the moon before crashing on its surface. They were the first of the unmanned space probes—Surveyor and Lunar Orbiter were later ones—to provide vital planning information about the lunar surface for the Apollo manned lunar landing program.

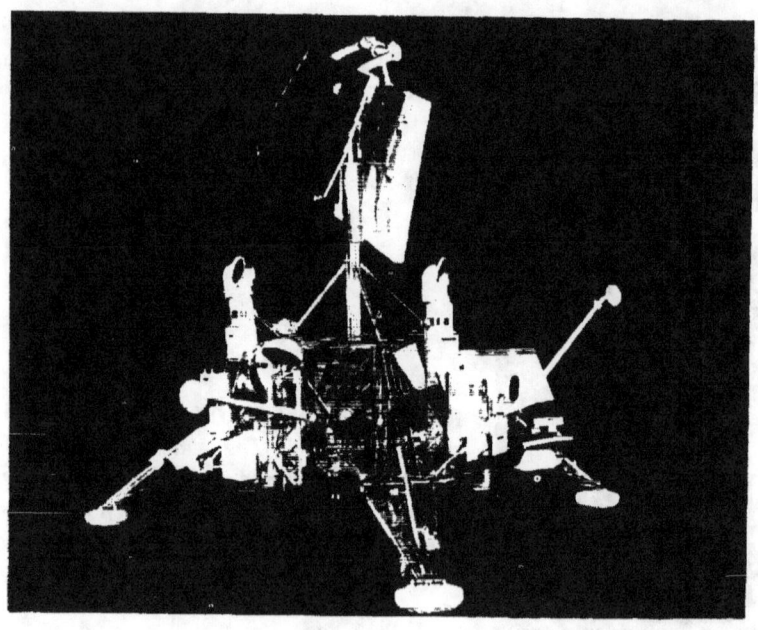

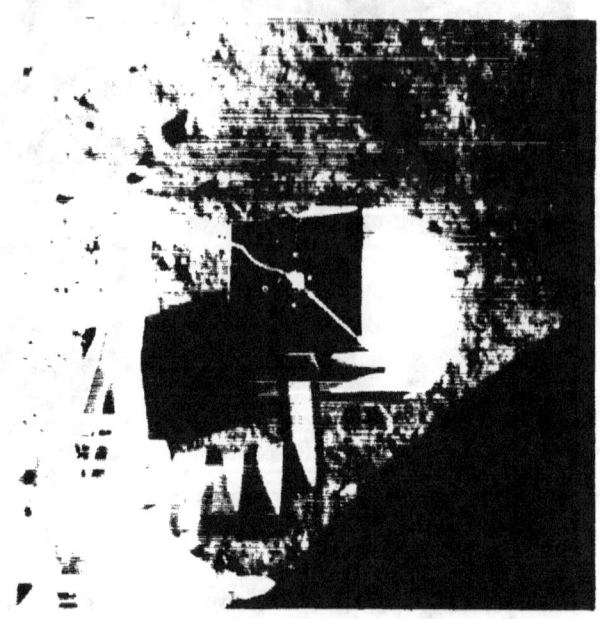

The Surveyor spacecraft, designed to make a softlanding on the moon. Surveyor 5's alpha-backscattering instrument, in the lower photo, analyzed chemical composition of the lunar surface after the 10 September 1967 landing.

SPACE PROBES

SURVEYOR. "Surveyor" was chosen in May 1960 to designate an advanced spacecraft series to explore and analyze the moon's surface. The designation was in keeping with the policy of naming lunar probes after "land exploration activities" established under the Cortright system of naming space probes.[1] Following the Ranger photographic lunar hardlanders, Surveyor probes marked an important advance in space technology: a softlanding on the moon's surface to survey it with television cameras and analyze its characteristics using scientific instruments.

Five Surveyor spacecraft—*Surveyor 1* in 1966; *Surveyor 3, 5,* and *6* in 1967; and *Surveyor 7* in 1968—softlanded on the moon and operated on the lunar surface over a combined time of approximately 17 months. They transmitted more than 87 000 photographs and made chemical and mechanical analyses of surface and subsurface samples.[2]

VIKING. The name "Viking" designated the planned first U.S. softlanding probes of the planet Mars.* The successor to Project Voyager, which was

Viking spacecraft model in simulated flight.

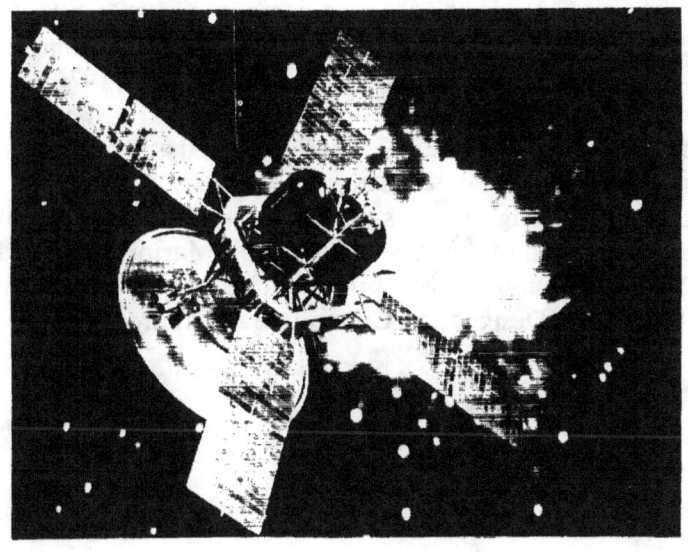

---

*"Viking had been previously used in the U.S. as the name for the early single-stage sounding rocket that later became the prototype for the first stage of the Vanguard launch vehicle. See Milton W. Rosen, *The Viking Rocket Story* (London: Faber and Faber, 1956) and Constance McL. Green and Milton Lomask, *Vanguard—A History,* NASA SP–4202 (Washington: NASA, 1970).

An artist's conception of the Viking Mars lander nearing touchdown on the Martian surface at Chryse. The parachute in the left background carries the aeroshell from which the lander detaches.

canceled in 1968, the Viking program was to send two unmanned spacecraft—each consisting of an orbiter and lander—to make detailed scientific measurements of the Martian surface, and search for indications of life forms.* The two Viking spacecraft, planned for launch in 1975 on Titan III–Centaur launch vehicles, were to reach Mars in 1976.

The name had been suggested by Walter Jacobowski in the Planetary Programs Office at NASA Headquarters and discussed at a management review held at Langley Research Center in November 1968.[1] It was the consensus at the meeting that "Viking" was a suitable name in that it reflected the spirit of nautical exploration in the same manner as "Mariner," according to the Cortright system of naming space probes.[2] The name was subsequently sent to the NASA Project Designation Committee and approved.

---

*Project Voyager was terminated because of the projected high cost of the program ($2.4 billion), which was related to the planned use of Saturn V launch vehicles.

IV

MANNED SPACE FLIGHT

View of the moon from Apollo 8.

MANNED SPACE FLIGHT

NASA's first four manned spaceflight projects were Mercury, Gemini, Apollo, and Skylab. As the first U.S. manned spaceflight project, Project Mercury—which included two manned suborbital flights and four orbital flights—"fostered Project Apollo and fathered Project Gemini." [1] The second manned spaceflight project initiated was the Apollo manned lunar exploration program. The national goal of a manned lunar landing in the 1960s was set forth by President John F. Kennedy 25 May 1961:

> . . . I believe that this nation should commit itself to achieving the goals, before this decade is out, of landing a man on the moon and returning him safely to earth. No single space project in this period will be more impressive to mankind, or more important for the long-range exploration of space; and none will be so difficult or expensive to accomplish. . . . But in a very real sense, it will not be one man going to the moon—if we make this judgment affirmatively, it will be an entire nation. [2]

The interim Project Gemini, completed in 1966, was conducted to provide spaceflight experience, techniques, and training in preparation for the complexities of Apollo lunar-landing missions. Project Skylab was originally conceived as a program to use hardware developed for Project Apollo in related manned spaceflight missions; it evolved into the Orbital Workshop program with three record-breaking missions in 1973–1974 to man the laboratory in earth orbit, producing new data on the sun, earth resources, materials technology, and effects of space on man.

The Apollo-Soyuz Test Project was an icebreaking effort in international cooperation. The United States and the U.S.S.R. were to fly a joint mission in 1975 to test new systems that permitted their spacecraft to dock with each other in orbit, for space rescue or joint research.

As technology and experience broadened man's ability to explore and use space, post-Apollo planning called for ways to make access to space more practical, more economical, nearer to routine. Early advanced studies grew into the Space Shuttle program. Development of the reusable space transportation system, to be used for most of the Nation's manned and unmanned missions in the 1980s, became the major focus of NASA's program for the 1970s. European nations cooperated by undertaking development of Spacelab, a pressurized, reusable laboratory to be flown in the Shuttle.

Apollo 11 *command and service module being readied for transport to the Vehicle Assembly Building at Kennedy Space Center, in left photo.* Apollo 11 *Astronaut Edwin E. Aldrin, Jr., below, setting up an experiment on the moon next to the lunar module. Opposite: the Greek god Apollo (courtesy of George Washington University).*

APOLLO. In July 1960 NASA was preparing to implement its long-range plan beyond Project Mercury and to introduce a manned circumlunar mission project—then unnamed—at the NASA/Industry Program Plans Conference in Washington. Abe Silverstein, Director of Space Flight Development, proposed the name "Apollo" because it was the name of a god in ancient Greek mythology with attractive connotations and the precedent for naming manned spaceflight projects for mythological gods and heroes had been set with Mercury.[1] Apollo was god of archery, prophecy, poetry, and music, and most significantly he was god of the sun. In his horse-drawn golden chariot, Apollo pulled the sun in its course across the sky each day.[2] NASA approved the name and publicly announced "Project Apollo" at the July 28–29 conference.[3]

Project Apollo took new form when the goal of a manned lunar landing was proposed to the Congress by President John F. Kennedy 25 May 1961 and was subsequently approved by the Congress. It was a program of three-man flights, leading to the landing of men on the moon. Rendezvous and docking in lunar orbit of Apollo spacecraft components were vital techniques for the intricate flight to and return from the moon.

The Apollo spacecraft consisted of the command module, serving as the crew's quarters and flight control section; the service module, containing propulsion and spacecraft support systems; and the lunar module, carrying

two crewmen to the lunar surface, supporting them on the moon, and returning them to the command and service module in lunar orbit. Module
designations came into use in 1962, when NASA made basic decisions on the
flight mode (lunar orbit rendezvous), the boosters, and the spacecraft for
Project Apollo. From that time until June 1966, the lunar module was called
"lunar excursion module (LEM)." It was renamed by the NASA Project
Designation Committee because the word "excursion" implied mobility on
the moon and this vehicle did not have that capability.[4] The later Apollo
flights, beginning with *Apollo 15,* carried the lunar roving vehicle (LRV), or
"Rover," to provide greater mobility for the astronauts while on the surface
of the moon.

Beginning with the flight of *Apollo 9,* code names for both the command
and service module (CSM) and lunar module (LM) were chosen by the
astronauts who were to fly on each mission. The code names were: *Apollo
9*—"Gumdrop" (CSM), "Spider" (LM); *Apollo 10*—"Charlie Brown"
(CSM), "Snoopy" (LM); *Apollo 11*—"Columbia" (CSM), "Eagle" (LM);
Apollo 12—"Yankee Clipper" (CSM), "Intrepid" (LM); *Apollo
13*—"Odyssey" (CSM), "Aquarius" (LM); *Apollo 14*—"Kitty Hawk"
(CSM), "Antares" (LM); *Apollo 15*—"Endeavour" (CSM), "Falcon"
(LM); *Apollo 16*—"Casper" (CSM), "Orion" (LM); *Apollo
17*—"America" (CSM); "Challenger" (LM).

The formula for numbering Apollo missions was altered when the three
astronauts scheduled for the first manned flight lost their lives in a flash fire
during launch rehearsal 27 January 1967. In honor of Astronauts Virgil I.
Grissom, Edward H. White II, and Roger B. Chaffee, the planned mission
was given the name "Apollo 1" although it was not launched. Carrying the
prelaunch designation AS–204 for the fourth launch in the Apollo Saturn IB
series, the mission was officially recorded as "First manned Apollo Saturn
flight—failed on ground test."

Manned Spacecraft Center Deputy Director George M. Low had urged
consideration of the request from the astronauts' widows that the designation "Apollo 1"—used by the astronauts publicly and included on their
insignia—be retained. NASA Headquarters Office of Manned Space Flight
therefore recommended the new numbering, and the NASA Project Designation Committee announced approval 3 April 1967.

The earlier, unmanned Apollo Saturn IB missions AS–201, AS–202, and
AS–203 were not given "Apollo" flight numbers and no missions were
named "Apollo 2" and "Apollo 3." The next mission flown, the first
Saturn V flight (AS–501, for Apollo Saturn V No. 1), skipped numbers 2

Lunar Rover parked on the moon during the Apollo 15 *mission.*

and 3 to become *Apollo 4* after launch into orbit 9 November 1967. Subsequent flights continued the sequence through *17*.[5]

The Apollo program carried the first men beyond the earth's field of gravity and around the moon on *Apollo 8* in December 1968 and landed the first men on the moon in *Apollo 11* on 20 July 1969. The program concluded with *Apollo 17* in December 1972 after putting 27 men into lunar orbit and 12 of them on the surface of the moon. Data, photos, and lunar samples brought to earth by the astronauts and data from experiments they left on the moon—still transmitting data in 1974—began to give a picture of the moon's origin and nature, contributing to understanding of how the earth had evolved.

APOLLO-SOYUZ TEST PROJECT (ASTP). The first international manned space project, the joint U.S.–U.S.S.R. rendezvous and docking mission took its name from the spacecraft to be used, the American Apollo and the Soviet Soyuz.

On 15 September 1969, two months after the *Apollo 11* lunar landing mission, the President's Space Task Group made its recommendations on the future U.S. space program. One objective was broad international par-

The Apollo spacecraft approaches the Soyuz for docking in orbit, in the artist's conception at top. Cosmonaut Aleksey A. Leonov and Astronaut Donald K. Slayton check out the docking module in a 1974 training session.

ticipation, and President Nixon included this goal in his March 1970 Space Policy Statement. The President earlier had approved NASA plans for increasing international cooperation in an informal meeting with Secretary of State William P. Rogers, Presidential Assistant for National Security Affairs Henry A. Kissinger, and NASA Administrator Thomas O. Paine aboard *Air Force One* while flying to the July *Apollo 11* splashdown.[1]

The United States had invited the U.S.S.R. to participate in experiments and information exchange over the past years. Now Dr. Paine sent Soviet Academy of Sciences President Mstislav V. Keldysh a copy of the U.S. post-Apollo plans and suggested exploration of cooperative programs. In April 1970 Dr. Paine suggested, in an informal meeting with Academician Anatoly A. Blagonravov in New York, that the two nations cooperate on astronaut safety, including compatible docking equipment on space stations and shuttles to permit rescue operations in space emergencies. Further discussions led to a 28 October 1970 agreement on joint efforts to design compatible docking arrangements. Three working groups were set up. Agreements on further details were reached in Houston, Texas, 21–25 June 1971 and in Moscow 29 November–6 December 1971. NASA Deputy Administrator George M. Low and a delegation met with a Soviet delegation in Moscow 4–6 April 1972 to draw up a plan for docking a U.S. Apollo spacecraft with a Russian Soyuz in earth orbit in 1975.[2]

Final official approval came in Moscow on 24 May 1972. U.S. President Nixon and U.S.S.R. Premier Aleksey N. Kosygin signed the Agreement Concerning Cooperation in the Exploration and Use of Outer Space for Peaceful Purposes, including development of compatible spacecraft docking systems to improve safety of manned space flight and to make joint scientific experiments possible. The first flight to test the systems was to be in 1975, with modified Apollo and Soyuz spacecraft. Beyond this mission, future manned spacecraft of the two nations would be able to dock with each other.[3]

During work that followed, engineers at Manned Spacecraft Center (renamed Johnson Space Center in 1973) shortened the lengthy "joint rendezvous and docking mission" to "Rendock," as a handy project name. But the NASA Project Designation Committee in June 1972 approved the official designation as "Apollo Soyuz Test Project (ASTP)," incorporating the names of the U.S. and U.S.S.R. spacecraft. The designation was sometimes written "Apollo/Soyuz Test Project," but the form "Apollo-Soyuz Test Project" was eventually adopted. NASA and the Soviet Academy of Sciences announced the official ASTP emblem in March 1974. The circular emblem displayed the English word "Apollo" and the Russian word "Soyuz" on either side of a center globe with a superimposed silhouette of the docked spacecraft.[4]

Scheduled for July 1975, the first international manned space mission would carry out experiments with astronauts and cosmonauts working together, in addition to testing the new docking systems and procedures. A three-module, two-man Soviet Soyuz was to be launched from the U.S.S.R.'s Baykonur Cosmodrome near Tyuratam on 15 July. Some hours later the modified Apollo command and service module with added docking module and a three-man crew would lift off on the Apollo-Skylab Saturn IB launch vehicle from Kennedy Space Center, to link up with the Soyuz. The cylindrical docking module would serve as an airlock for transfer of crewmen between the different atmospheres of the two spacecraft. After two days of flying joined in orbit, with crews working together, the spacecraft would undock for separate activities before returning to the earth.[5]

GEMINI. In 1961 planning was begun on an earth-orbital rendezvous program to follow the Mercury project and prepare for Apollo missions. The improved or "Advanced Mercury" concept was designated "Mercury Mark II" by Glenn F. Bailey, NASA Space Task Group Contracting Officer, and John Y. Brown of McDonnell Aircraft Corporation.[1] The two-man spacecraft was based on the one-man Mercury capsule, enlarged and made capable of longer flights. Its major purposes were to develop the technique of rendezvous in space with another spacecraft and to extend orbital flight time.

NASA Headquarters personnel were asked for proposals for an appropriate name for the project and, in a December 1961 speech at the Industrial College of the Armed Forces, Dr. Robert C. Seamans, Jr., then NASA Associate Administrator, described Mercury Mark II, adding an offer of a token reward to the person suggesting the name finally accepted. A member of the audience sent him the name "Gemini." Meanwhile, Alex P. Nagy in NASA's Office of Manned Space Flight also had proposed "Gemini." Dr. Seamans recognized both as authors of the name.[2]

"Gemini," meaning "twins" in Latin, was the name of the third constellation of the zodiac, made up of the twin stars Castor and Pollux. To Nagy it seemed an appropriate connotation for the two-man crew, a rendezvous mission, and the project's relationship to Mercury. Another connotation of the mythological twins was that they were considered to be the patron gods of voyagers.[3] The nomination was selected from several made in NASA Headquarters, including "Diana," "Valiant," and "Orpheus"

The Gemini 7 spacecraft was photographed from the window of Gemini 6 during rendezvous maneuvers 15 December 1965. Castor and Pollux, the Gemini of mythology, ride their horses through the sky (courtesy of the Library of Congress).

from the Office of Manned Space Flight. On 3 January 1962, NASA announced the Mercury Mark II project had been named "Gemini."[4]

After 12 missions—2 unmanned and 10 manned—Project Gemini ended 15 November 1966. Its achievements had included long-duration space flight, rendezvous and docking of two spacecraft in earth orbit, extravehicular activity, and precision-controlled reentry and landing of spacecraft.

The crew of the first manned Gemini mission, Astronauts Virgil I. Grissom and John W. Young, nicknamed their spacecraft "Molly Brown." The name came from the musical comedy title, *The Unsinkable Molly Brown,* and was a facetious reference to the sinking of Grissom's Mercury-

Redstone spacecraft after splashdown in the Atlantic Ocean 21 July 1961. "Molly Brown" was the last Gemini spacecraft with a nickname; after the *Gemini 3* mission, NASA announced that "all Gemini flights should use as official spacecraft nomenclature a single easily remembered and pronounced name." [5]

Astronaut Edward H. White II floats in space, secured to the Gemini 4 *spacecraft.*

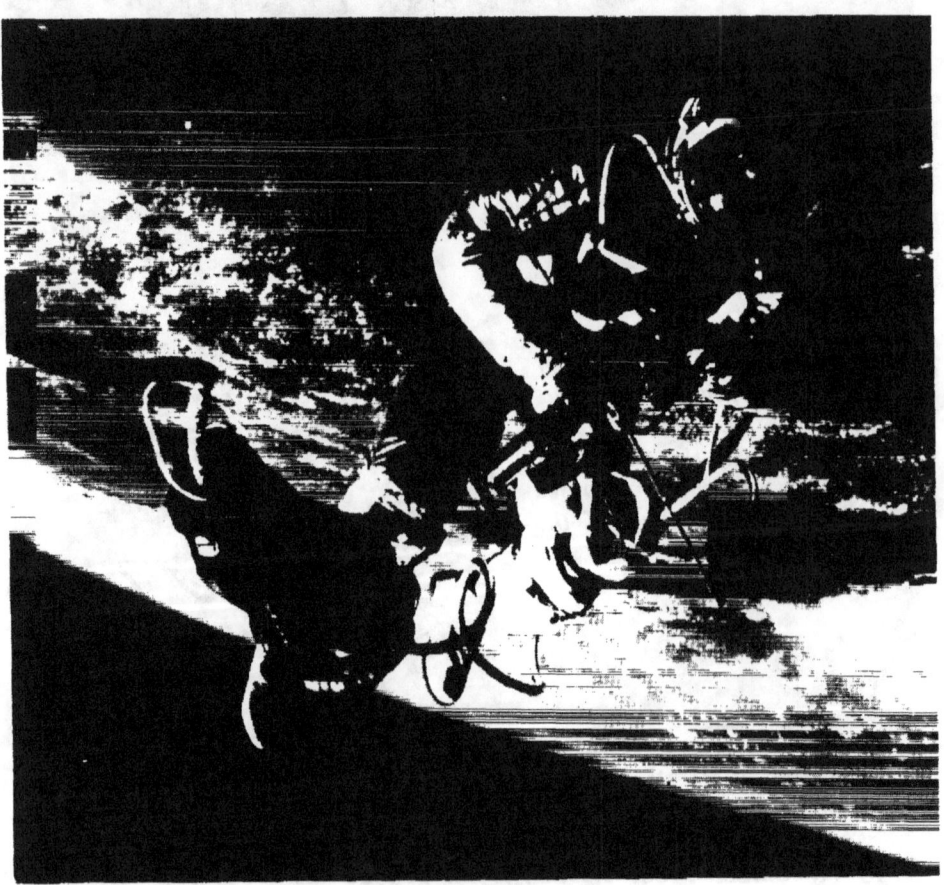

MERCURY. Traditionally depicted wearing a winged cap and winged shoes, Mercury was the messenger of the gods in ancient Roman and (as Hermes) Greek mythology. [1] The symbolic associations of this name appealed to Abe Silverstein, NASA's Director of Space Flight Development, who suggested it for the manned spaceflight project in the autumn of 1958. On 26 November 1958 Dr. T. Keith Glennan, NASA Administrator, and Dr. Hugh

Full-scale mockups of the Mercury and Gemini spacecraft.

L. Dryden, Deputy Administrator, agreed upon "Mercury," and on 17 December 1958 Dr. Glennan announced the name for the first time.[2]

On 9 April 1959 NASA announced selection of the seven men chosen to be the first U.S. space travelers, "astronauts." The term followed the semantic tradition begun with "Argonauts," the legendary Greeks who traveled far and wide in search of the Golden Fleece, and continued with "aeronauts"—pioneers of balloon flight.[3] Robert R. Gilruth, head of the Space Task Group, proposed "Project Astronaut" to NASA Headquarters, but the suggestion lost out in favor of Project Mercury "largely because it [Project Astronaut] might lead to overemphasis on the personality of the man."[4]

In Project Mercury the United States acquired its first experience in conducting manned space missions and its first scientific and engineering knowledge of man in space. After two suborbital and three orbital missions, Project Mercury ended with a fourth orbital space flight—a full-day mission by L. Gordon Cooper, Jr., 15–16 May 1963.

In each of Project Mercury's manned space flights, the assigned astronaut chose a call sign for his spacecraft just before his mission. The choice of

"Freedom 7" by Alan B. Shepard, Jr., established the tradition of the numeral "7," which came to be associated with the team of seven Mercury astronauts. When Shepard chose "Freedom 7," the numeral seemed significant to him because it appeared that "capsule No. 7 on booster No. 7 should be the first combination of a series of at least seven flights to put Americans into space." [5] The prime astronaut for the second manned flight, Virgil I. Grissom, named his spacecraft "Liberty Bell 7" because "the name was to Americans almost synonymous with 'freedom' and symbolical numerically of the continuous teamwork it represented." [6]

John Glenn, assigned to take the Nation's first orbital flight, named his Mercury spacecraft "Friendship 7." Scott Carpenter chose "Aurora 7," he said, "because I think of Project Mercury and the open manner in which we are conducting it for the benefit of all as a light in the sky. Aurora also

Astronaut John H. Glenn, Jr., is hoisted out of the Friendship 7 *spacecraft after splashdown in the Atlantic 20 February 1962. The god Mercury, poised for flight, at right (courtesy of the National Gallery of Art).*

means dawn—in this case the dawn of a new age. The 7, of course, stands for the original seven astronauts." [7] Walter M. Schirra selected "Sigma 7" for what was primarily an engineering flight—a mission to evaluate spacecraft systems; "sigma" is an engineering symbol for summation. In selecting "sigma," Schirra also honored "the immensity of the engineering effort behind him." [8] Cooper's choice of "Faith 7" symbolized, in his words, "my trust in God, my country, and my teammates." [9]

SKYLAB. Planning for post-Apollo manned spaceflight missions evolved directly from the capability produced by the Apollo and Saturn technologies, and Project Skylab resulted from the combination of selected program objectives. In 1964, design and feasibility studies had been initiated for missions that could use modified Apollo hardware for a number of possible lunar and earth-orbital scientific and applications missions. The study concepts were variously known as "Extended Apollo (Apollo X)" and the "Apollo Extension System (AES)." [1] In 1965 the program was coordinated under the name "Apollo Applications Program (AAP)" and by 1966 had narrowed in scope to primarily an earth-orbital concept. [2]

Projected AAP missions included the use of the Apollo Telescope Mount (ATM). In one plan it was to be launched separately and docked with an orbiting workshop in the "wet" workshop configuration. The wet workshop—using the spent S–IVB stage of the Saturn I launch vehicle as a workshop after purging it in orbit of excess fuel—was later dropped in favor of the "dry" configuration using the Saturn V launch vehicle. The extra fuel carried by the S–IVB when used as a third stage on the Saturn V, for moon launches, would not be required for the Skylab mission, and the stage could be completely outfitted as a workshop before launch, including the ATM. [3]

The name "Skylab," a contraction connoting "laboratory in the sky," was suggested by L/C Donald L. Steelman (USAF) while assigned to NASA. He later received a token reward for his suggestion. Although the name was proposed in mid-1968, NASA decided to postpone renaming the program because of budgetary considerations. "Skylab" was later referred to the NASA Project Designation Committee and was approved 17 February 1970. [4]

Skylab 1 (SL–1), the Orbital Workshop with its Apollo Telescope Mount, was put into orbit 14 May 1973. Dynamic forces ripped off the meteoroid shield and one solar array wing during launch, endangering the entire program, but the three astronauts launched on *Skylab 2* (SL–2)—the first manned mission to crew the Workshop—were able to repair the spacecraft and completed 28 days living and working in space before their safe return.

Skylab Orbital Workshop photographed from the Skylab 2 *command module during fly-around inspection. The Workshop's remaining solar array wing, after the second wing was ripped off during launch, is deployed below the ATM's four arrays. The emergency solar parasol erected by the astronauts is visible on the lower part of the spacecraft. The cutaway drawing shows crew quarters and work areas.*

They were followed by two more three-man crews during 1973. The *Skylab 3* crew spent 59 days in space and *Skylab 4* spent 84. Each Skylab mission was the longest-duration manned space flight to that date, also setting distance-in-orbit and extravehicular records. *Skylab 4,* the final mission (16 November 1973 to 8 February 1974) recorded the longest in-orbit EVA (7 hours 1 minute), the longest cumulative orbital EVA time for one mission (22 hours 21 min in four EVAs), and the longest distance in orbit for a manned mission (55.5 million kilometers).

The Skylab missions proved that man could live and work in space for extended periods; expanded solar astronomy beyond earth-based observations, collecting new data that could revise understanding of the sun and its effects on the earth; and returned much information from surveys of earth resources with new techniques. The deactivated Workshop remained in orbit; it might be visited by a future manned flight, but was not to be inhabited again.

SPACE SHUTTLE. The name "Space Shuttle" evolved from descriptive references in the press, aerospace industry, and Government and gradually came into use as concepts of reusable space transportation developed. As early NASA advanced studies grew into a full program, the name came into official use.* [1]

From its establishment in 1958, NASA studied aspects of reusable launch vehicles and spacecraft that could return to the earth. The predecessor National Advisory Committee for Aeronautics and then NASA cooperated with the Air Force in the X–15 rocket research aircraft program in the 1950s and 1960s and in the 1958–1963 Dyna-Soar ("Dynamic-Soaring") hypersonic boost-glide vehicle program. Beginning in 1963, NASA joined the USAF in research toward the Aerospaceplane, a manned vehicle to go into orbit and return, taking off and landing horizontally. Joint flight tests in the 1950s and 1960s of wingless lifting bodies—the M2 series, HL–10, and eventually the X–24—tested principles for future spacecraft reentering the atmosphere.

Marshall Space Flight Center sponsored studies of recovery and reuse of the Saturn V launch vehicle. MSFC Director of Future Projects Heinz H. Koelle in 1962 projected a "commercial space line to earth orbit and the

---

*In January 1975, NASA's Project Designation Committee was considering suggestions for a new name for the Space Shuttle, submitted by Headquarters and Center personnel and others at the request of Dr. George M. Low, NASA Deputy Administrator. Rockwell International Corporation, Shuttle prime contractor, was reported as referring to it as "Spaceplane." (Bernice M. Taylor, Administrative Assistant to Assistant Administrator for Public Affairs, NASA, telephone interview, 12 Feb. 1975; and *Aviation Week & Space Technology,* 102 [20 January 1975], 10.)

The Space Shuttle lifts off in the artist's conception of missions of the 1980s, at left, with booster jettison and tank jettison following in sequence, as the orbiter heads for orbit and its mission.

moon," for cargo transportation by 1980 or 1990. Leonard M. Tinnan of MSFC published a 1963 description of a winged, flyback Saturn V.[2] Other studies of "logistics spacecraft systems," "orbital carrier vehicles," and "reusable orbital transports" followed throughout the 1960s in NASA, the Department of Defense, and industry.

MANNED SPACE FLIGHT

As the Apollo program neared its goal, NASA's space program objectives widened and the need for a fully reusable, economical space transportation system for both manned and unmanned missions became more urgent. In 1966 the NASA budget briefing outlined an FY 1967 program including advanced studies of "ferry and logistics vehicles." The President's Science Advisory Committee in February 1967 recommended studies of more economical ferry systems with total recovery and rescue possibilities.[3] Industry studies under NASA contracts 1969–1971 led to definition of a reusable Space Shuttle system and to a 1972 decision to develop the Shuttle.

The term "shuttle" crept into forecasts of space transportation at least as early as 1952. In a *Collier's* article, Dr. Wernher von Braun, then Director of the U.S. Army Ordnance Guided Missiles Development Group, envisioned space stations supplied by rocket ships that would enter orbit and return to earth to land "like a normal airplane," with small, rocket-powered "shuttle-craft," or "space taxis," to ferry men and materials between rocket ship and space station.[4]

In October 1959 Lockheed Aircraft Corporation and Hughes Aircraft Company reported plans for a space ferry or "commuter express," for "shuttling" men and materials between earth and outer space. In December, *Christian Science Monitor* Correspondent Courtney Sheldon wrote of the future possibility of a "man-carrying space shuttle to the nearest planets."[5]

The term reappeared occasionally in studies through the early 1960s. A 1963 NASA contract to Douglas Aircraft Company was to produce a conceptual design for Philip Bono's "Reusable Orbital Module Booster and Utility Shuttle (ROMBUS)," to orbit and return to touch down with legs like the lunar landing module's. Jettison of eight strap-on hydrogen tanks for recovery and reuse was part of the concept.[6] The press—in accounts of European discussions of Space Transporter proposals and in articles on the Aerospaceplane, NASA contract studies, USAF START reentry studies, and the joint lifting-body flights—referred to "shuttle" service, "reusable orbital shuttle transport," and "space shuttle" forerunners. *

In 1965 Dr. Walter R. Dornberger, Vice President for Research of Textron Corporation's Bell Aerosystems Company, published "Space Shuttle of the Future: The Aerospaceplane" in Bell's periodical *Rendezvous*. In July Dr. Dornberger gave the main address in a University of Tennessee Space Institute short course: "The Recoverable, Reusable Space Shuttle." [7]

NASA used the term "shuttle" for its reusable transportation concept officially in 1968. Associate Administrator for Manned Space Flight George E. Mueller briefed the British Interplanetary Society in London in August with charts and drawings of "space shuttle" operations and concepts. In November, addressing the National Space Club in Washington, D.C., Dr. Mueller declared the next major thrust in space should be the space shuttle.[8]

By 1969 "Space Shuttle" was the standard NASA designation, although some efforts were made to find another name as studies were pursued.[9] The "Space Shuttle" was given an agency-wide code number; the Space Shuttle Steering Group and Space Shuttle Task Group were established. In September the Space Task Group appointed by President Nixon to help define post-Apollo space objectives recommended the U.S. develop a reusable, economic space transportation system including a shuttle. And in October feasibility study results were presented at a Space Shuttle Conference in Washington. Intensive design, technology, and cost studies followed in 1970 and 1971.[10]

---

*The *Defense/Space Business Daily* newsletter was persistent in referring to USAF and NASA reentry and lifting-body tests as "Space Shuttle" tests. Editor-in-Chief Norman L. Baker said the newsletter had first tried to reduce the name "Aerospaceplane" to "Spaceplane" for that project and had moved from that to "Space Shuttle" for reusable, back-and-forth space transport concepts as early as 1963. The name was suggested to him by the Washington, D.C., to New York airline shuttle flights. (Telephone interview, 22 April 1975.)

Application of the word "shuttle" to anything that moved quickly back and forth (from shuttlecock to shuttle train and the verb "to shuttle") had arisen in the English language from the name of the weaving instrument that passed or "shot" the thread of the woof from one edge of the cloth to the other. The English word came from the Anglo-Saxon "scytel" for missile, related to the Danish "skyttel" for shuttle, the Old Norwegian "skutill" for harpoon, and the English "shoot." (*Webster's International Dictionary*, ed. 2, unabridged.)

On 5 January 1972 President Nixon announced that the United States would develop the Space Shuttle.

The Space Shuttle would be a delta-winged aircraftlike orbiter about the size of a DC-9 aircraft, mounted at launch on a large, expendable liquid-propellant tank and two recoverable and reusable solid-propellant rocket boosters (SRBs) that would drop away in flight. The Shuttle's cargo bay eventually would carry most of the Nation's civilian and military payloads. Each Shuttle was to have a lifetime of 100 space missions, carrying up to 29 500 kilograms at a time. Sixty or seventy flights a year were expected in the 1980s.

Flown by a three-man crew, the Shuttle would carry satellites to orbit, repair them in orbit, and later return them to earth for refurbishment and reuse. It would also carry up to four scientists and engineers to work in a pressurized laboratory (see Spacelab) or technicians to service satellites. After a 7- to 30-day mission, the orbiter would return to earth and land like an aircraft, for preparation for the next flight.

At the end of 1974, parts were being fabricated, assembled, and tested for flight vehicles. Horizontal tests were to begin in 1977 and orbital tests in 1979. The first manned orbital flight was scheduled for March 1979 and the complete vehicle was to be operational in 1980.

SPACE TUG. Missions to orbits higher than 800 kilometers would require an additional propulsion stage for the Space Shuttle. A reusable "Space Tug" would fit into the cargo bay to deploy and retrieve payloads beyond the orbiter's reach and to achieve earth-escape speeds for deep-space exploration. Under a NASA and Department of Defense agreement, the Air Force was to develop an interim version—the "interim upper stage (IUS)," named by the Air Force the "orbit-to-orbit stage (OOS)," to be available in 1980. NASA meanwhile continued planning and studies for a later full-capacity Space Tug.[11]

Joseph E. McGolrick of the NASA Office of Launch Vehicles had used the term in a 1961 memorandum suggesting that, as capabilities and business in space increased, a need might arise for "a space tug—a space vehicle capable of orbital rendezvous and . . . of imparting velocities to other bodies in space." He foresaw a number of uses for such a vehicle and suggested it be considered with other concepts for the period after 1970. McGolrick thought of the space tug as an all-purpose workhorse, like the small, powerful tugboats that moved huge ocean liners and other craft. The name was used frequently in studies and proposals through the years, and in September 1969 the Presidential Space Task Group's recommendation for a new space transportation system proposed development of a reusable, chemically propelled space tug, as well as a shuttle and a nuclear stage.[12]

LARGE SPACE TELESCOPE. Among Shuttle payloads planned—besides Spacelab and satellites like those launched in the past by expendable boosters—was the Large Space Telescope (LST), to be delivered to orbit as an international facility for in-orbit research controlled by scientists on the ground. The LST would observe the solar system and far galaxies from above the earth's atmosphere. On revisits, the Shuttle would service the orbiting telescope, exchange scientific hardware, and—several years later—return the LST to the earth.

LONG-DURATION EXPOSURE FACILITY. Another payload was to be placed in orbit for research into effects of exposure to space. The unmanned, free-flying Long-Duration Exposure Facility (LDEF) would expose a variety of passive experiments in orbit and would later be retrieved for refurbishment and reuse.

SPACELAB. A new venture in space flight made possible by the Space Shuttle, Spacelab was to be a reusable "space laboratory" in which scientists and engineers could work in earth orbit without spacesuits or extensive astronaut training. The program drew the United States and Europe into closer cooperation in space efforts.

The name finally chosen for the space laboratory was that used by the European developers. It followed several earlier names used as NASA's program developed toward its 1980s operational goal. In 1971 NASA awarded a contract for preliminary design of "Research and Applications Modules" (RAMs) to fly on the Space Shuttle. A family of manned or "man-tended" payload carriers, the RAMs were to provide versatile laboratory facilities for research and applications work in earth orbit. Later modules were expected to be attached to space stations, in addition to the earlier versions operating attached to the Shuttle. The simplest RAM mode was called a "Sortie Can" at Marshall Space Flight Center. It was a low-cost, simplified, pressurized laboratory to be carried on the Shuttle orbiter for short "sortie" missions into space.[1] In June 1971 the NASA Project Designation Committee redesignated the Sortie Can the "Sortie Lab," as a more fitting name.[2]

When the President's Space Task Group had originally recommended development of the Space Shuttle in 1969, it had also recommended broad international participation in the space program, and greater international cooperation was one of President Nixon's Space Policy Statement goals in March 1970. NASA Administrator Thomas O. Paine visited European

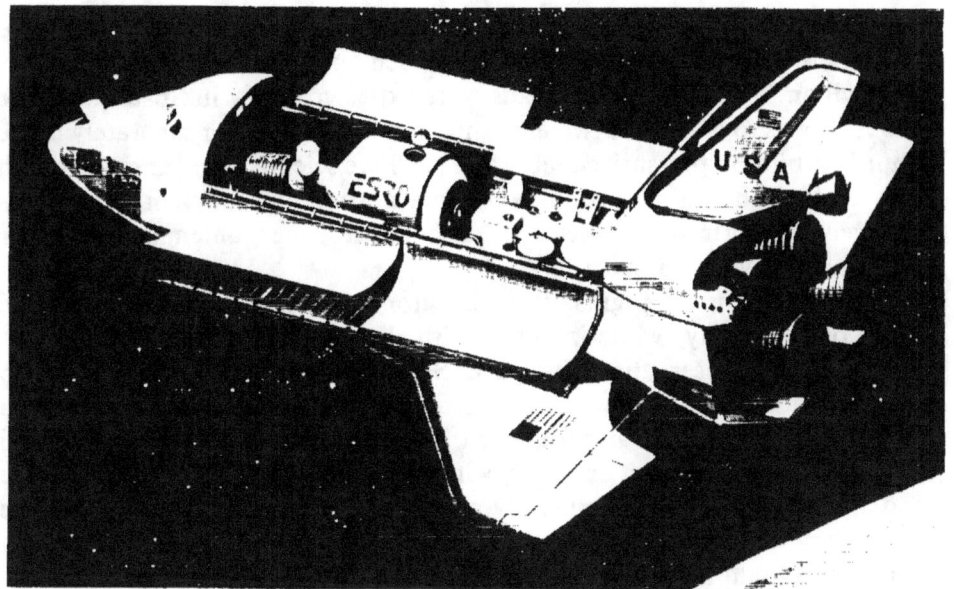

A Spacelab module and pallet fill the payload bay of a scale-model Space Shuttle orbiter. The laboratory module is nearest the cabin.

capitals in October 1969 to explain Shuttle plans and invite European interest, and 43 European representatives attended a Shuttle Conference in Washington. One area of consideration for European effort was development of the Sortie Lab.[3]

On 20 December 1972 a European Space Council ministerial meeting formally endorsed European Space Research Organization development of Sortie Lab. An intergovernmental agreement was signed 10 August 1973 and ESRO and NASA initialed a memorandum of understanding. The memorandum was signed 24 September 1973. Ten nations—Austria, Belgium, Denmark, France, West Germany, Italy, the Netherlands, Spain, Switzerland, and the United Kingdom—would develop and manufacture the units. The first unit was to be delivered to NASA free in the cooperative program, and NASA would buy additional units. NASA would fly Spacelab on the Shuttle in cooperative missions, in U.S. missions, and for other countries with costs reimbursed.[4]

In its planning and studies, ESRO called the laboratory "Spacelab." And when NASA and ESRO signed the September 1973 memorandum on cooperation NASA Administrator James C. Fletcher announced that NASA's Sortie Lab program was officially renamed "Spacelab," adopting the ESRO name.[5]

Spacelab was designed as a low-cost laboratory to be quickly available to users for a wide variety of orbital research and applications. Almost half the civilian Space Shuttle payloads were expected to fly in Spacelab in the 1980s. It was to consist of two elements, carried together or separately in the Shuttle orbiter: a pressurized laboratory, where scientists and engineers with only brief flight training could work in a normal environment, and an instrument platform, or "pallet," to support telescopes, antennas, and other equipment exposed to space.

Reusable for 50 flights, the laboratory would remain in the Shuttle hold, or cargo bay, while in orbit, with the bay doors held open for experiments and observations in space. Seven-man missions, many of them joint missions with U.S. and European crew members, would include a three-man Shuttle crew and four men for Spacelab. Up to three men could work in the laboratory at one time, with missions lasting 7 to 30 days. At the end of each flight, the orbiter would make a runway landing and the laboratory would be removed and prepared for its next flight. Racks of experiments would be prepared in the home laboratories on the ground, ready for installation in Spacelab for flight and then removal on return.[6]

One of the planned payloads was NASA's AMPS (Atmospheric, Magnetospheric, and Plasmas-in-Space) laboratory, to be installed in Spacelab for missions in space.[7]

At the end of 1974, life scientists, astronomers, atmospheric physicists, and materials scientists were defining experiment payloads for Spacelab. The first qualified flight unit was due for delivery in 1979 for 1980 flight. A European might be a member of the first flight crew.[8]

V

SOUNDING ROCKETS

High-gain antenna at Wallops Flight Center receives telemetry signals from experiments launched on sounding rockets.

SOUNDING ROCKETS

Sounding rockets are rockets that carry instruments into the upper atmosphere to investigate its nature and characteristics, gathering data from meteorological measurements at altitudes as low as 32 kilometers to data for ionospheric and cosmic physics at altitudes up to 6400 kilometers.[1] Sounding rockets also flight-test instruments to be used in satellites. The term "sounding rocket" derived from the analogy to maritime soundings made of the ocean depths.[2]

Sending measurement instruments into the high atmosphere was one of the principal motives for 20th century rocket development. This was the stated purpose of Dr. Robert H. Goddard in his rocket design studies as early as 1914.[3] But it was not until 1945 that the first U.S. Government-sponsored sounding rocket was launched—the Wac Corporal, a project of the Jet Propulsion Laboratory and U.S. Army Ordnance.[4]

Sounding rockets played an important role in the International Geophysical Year (IGY), an 18-month period (1 July 1957 to 31 December 1958) coinciding with high solar activity. The IGY was an intensive investigation of the natural environment—the earth, the oceans, and the atmosphere—by 30 000 participants representing 66 nations. More than 300 instrumented sounding rockets launched from sites around the world made significant discoveries regarding the atmosphere, the ionosphere, cosmic radiation, auroras, and geomagnetism.[5]

The International Years of the Quiet Sun (1 January 1964 to 31 December 1965), a full-scale follow-up to the IGY, was an intensive effort of geophysical observations in a period of minimum solar activity. Instrumented sounding rockets again played a significant role in the investigation of earth-sun interactions. By the end of 1974, some 20 countries had joined NASA in cooperative projects launching more than 1700 rockets from ranges in the United States and abroad.[6]

Sounding rocket research gave rise to three new branches of astronomy—ultraviolet, x-ray, and gamma ray. Experiments launched on rockets have characterized the main features of the earth's upper atmosphere and contributed the first recognition of the geocorona, knowledge of ionospheric chemistry, detection of electrical currents in the ionosphere, and de-

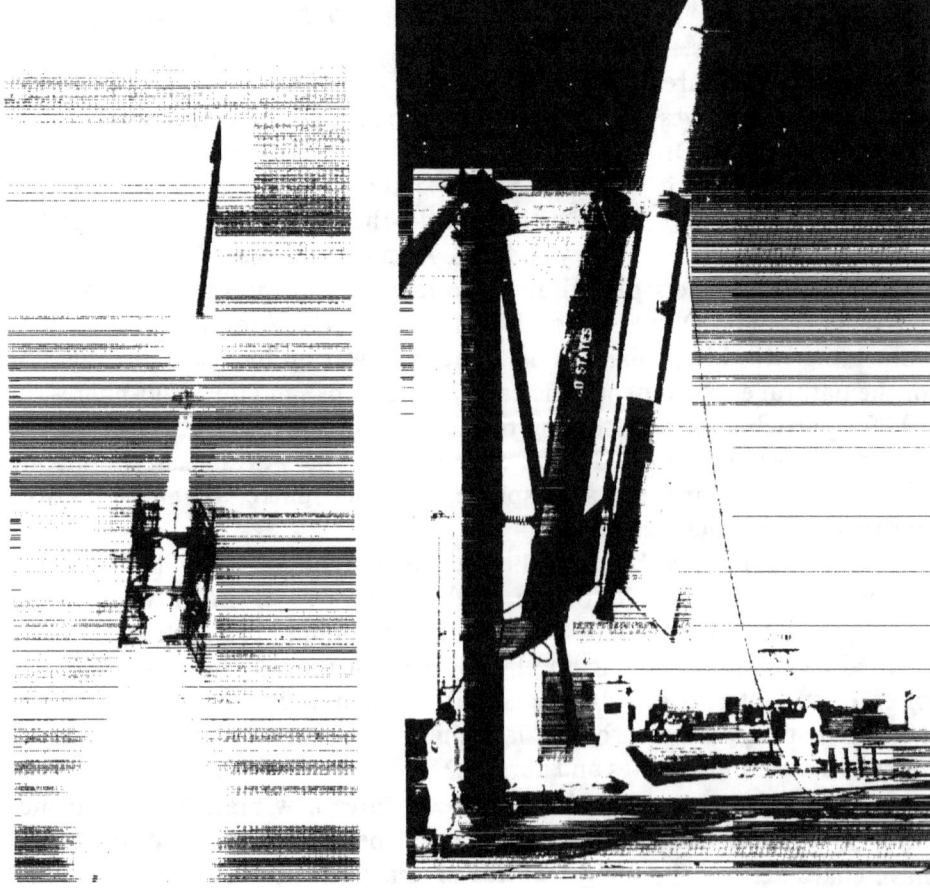

Aerobee 150A, top, in assembly area. At left below, Aerobee 350 launched on its first full flight test, 18 June 1965. At right below, Astrobee 1500 erected for its first flight test, 21 October 1964.

scription of particle flux in auroras. One of the earliest discoveries was of solar x-rays originating in the solar corona.[7]

Because higher performance sounding rockets were not economical for low-altitude experiments and lower performance rockets were not useful for high-altitude experiments, NASA used a number of rockets of varying capabilities—including Aerobee and Astrobee, Arcas, Argo D-4 (Javelin), Nike-Apache, Nike-Cajun, Nike-Hawk, Nike-Malemute, Nike-Tomahawk, Terrier-Malemute, and Black Brant. A high-performance rocket, the Aries, was under development in 1974. Vehicles could economically place 5 to 900 kilograms at altitudes up to 2200 kilometers. Highly accurate payload pointing and also payload recovery were possible when needed.[8]

AEROBEE, ASTROBEE. Development of the Aerobee liquid-propellant sounding rocket was begun in 1946 by the Aerojet Engineering Corporation (later Aerojet-General Corporation) under contract to the U.S. Navy. The Applied Physics Laboratory (APL) of Johns Hopkins University was assigned technical direction of the project. James A. Van Allen, then Director of the project at APL, proposed the name "Aerobee." He took the "Aero" from Aerojet Engineering and the "bee" from Bumblebee, the name of the overall project to develop naval rockets [1] that APL was monitoring for the Navy. The 18-kilonewton-thrust, two-stage Aerobee was designed to carry a 68-kilogram payload to a 130-kilometer altitude.

In 1952, at the request of the Air Force and the Navy, Aerojet undertook design and development of the Aerobee-Hi, a high-performance version of the Aerobee designed expressly for research in the upper atmosphere.[2] An improved Aerobee-Hi became the Aerobee 150. The uprated Aerobee 150 was named "Astrobee." Aerojet used the prefix "Aero" to designate liquid-propellant sounding rockets and "Astro" for its solid-fueled rockets.[3] Some of the Aerobee and Astrobee models employed by NASA were:[4]

Sounding Rocket	Payload Weight (kg)	Nominal Altitude* (km)
Aerobee 150 and 150A**	68	270
	227	110
Aerobee 170 and 170A	113.5	250
	227	150
Aerobee 200 and 200A	113.5	310
	227	290

Sounding Rocket	Payload Weight (kg)	Nominal Altitude* (km)
Aerobee 350	136	400
	454	210
Astrobee 1500	45	2200
	136	1200
Astrobee F............................	91	500
	454	400

*Sea level launch at 85° launch elevation.
**The "A" designation indicated the rocket had four fins instead of three.

APACHE. The Apache solid-propellant rocket stage was used with the Nike first stage. Identical in appearance to the Nike-Cajun, the Nike-Apache could reach higher altitudes because the Apache propellant burning time was longer (6.4 seconds versus Cajun's 4 seconds). It could carry 34-kilogram payloads to an operating altitude of 210 kilometers or 100 kilograms to 125 kilometers.[1]

Technicians ready a Nike-Apache on board the USNS Croatan, *Wallops Flight Center mobile range facility.*

124

The name "Apache," from the name of the American Indian tribe, followed Thiokol Chemical Corporation (later Thiokol Corporation) tradition of giving Thiokol-developed stages Indian-related names, which had begun with Cajun. [2]

ARCAS. A small solid-propellant sounding rocket, Arcas was named in 1959 by its producer, Atlantic Research Corporation. The name was an acronym for "All-purpose Rocket for Collecting Atmospheric Soundings." [1] It was intentional that the first three letters, "A–R–C," also were the initials of the Atlantic Research Corporation. [2] An inexpensive vehicle designed specifically for meteorological research, Arcas could carry a five-kilogram payload to an altitude of 64 kilometers. [3] Later versions were the Boosted Arcas, Boosted Arcas II, and Super Arcas, all of which NASA used.

Two other sounding rockets developed by Atlantic Research were used briefly by NASA. The Arcon was named by the Corporation and the Iris was named by Eleanor Pressly of Goddard Space Flight Center, which managed the rockets. [4]

Arcas being loaded into its launch tube.

ARGO. The name of a series of sounding rockets, "Argo" was from the name of Jason's ship in the ancient Greek myth of Jason's travels in search of the Golden Fleece.* The first sounding rocket in this series, developed by the Aerolab Company (later a division of Atlantic Research Corporation), was called "Jason." Subsequent vehicles in the series were given names also

---

* Jason named his ship after its builder, Argus. See Thomas Bulfinch,*Mythology,* Edmund Fuller, ed. (New York: Dell Publishing Co., Inc., 1959), p. 108.

Javelin in horizontal position on the launcher, for last-minute checks during prelaunch operations.

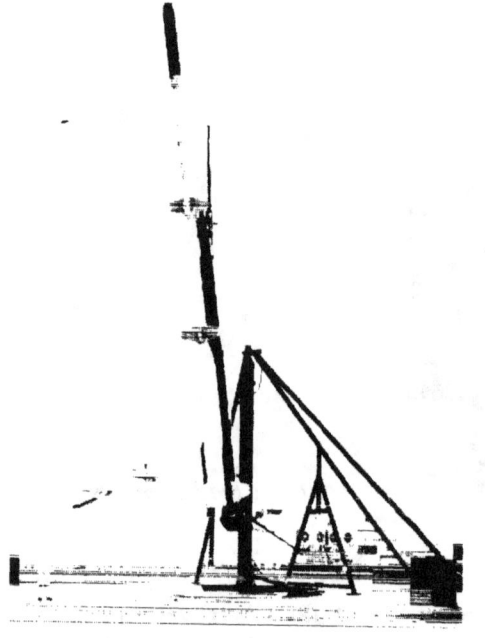

Journeyman

beginning with the letter "J": The Argo D–4 and Argo D–8 were named "Javelin" and "Journeyman." The "D–4" and "D–8" designations referred to the number of stages—"D" for "four"—and to the design revision—fourth and eighth.[1]

Argo D–4 (Javelin) was designed to carry 40- to 70-kilogram payloads to 800- to 1100-kilometer altitudes. Argo D–8 (Journeyman) could carry 20- to 70-kilogram payloads to 1500- to 2100-kilometer altitudes.[2] Javelin was still used by NASA in 1974, but Journeyman was discontinued in 1965.[3] Javelin was also mated to the Nike first stage for heavier payloads.

ARIES. NASA in 1974 was working with the Naval Research Laboratory, Sandia Laboratories, and West Germany to develop a new sounding rocket, the Aries, using surplus second stages from the Department of Defense Minuteman intercontinental ballistic missiles. The rocket, which had flown three-test flights by December 1974, would lift larger payloads for longer flight times than other rockets—in astronomy, physics, and space processing research projects.[1]

The Aries would have greater volume for carrying experiment instruments than provided by the Aerobee 350 sounding rocket and would carry 180- to 900-kilogram scientific payloads to altitudes that would permit 11 to 7 minutes viewing time above 91 440 meters, appreciably longer than the viewing time of the Aerobee 350 and the Black Brant VC. (The first test flights had carried 817 kilograms to 270.7 and 299 kilometers.)[2] It also was expected to give 11 to 8 minutes in weightless conditions for materials-processing-experiment payloads of 45 to 454 kilograms.[3]

When the project was first conceived, the new vehicle was called "Fat Albert" after the television cartoon character, because its short, fat appearance contrasted with that of other rockets. The Naval Research Laboratory asked Robert D. Arritt of its Space Science Division to choose a more dignified name. Arritt and a group of his colleagues chose "Aries"; it was the name of a constellation (the rocket would be used for astronomy projects) and it was "a name that was available." It also was Arritt's zodiac sign.[4]

ASP. The name of the solid-propellant sounding rocket "Asp" was an acronym for "Atmospheric Sounding Projectile." Designed to carry up to 36 kilograms of payload, the Asp was developed by Cooper Development Corporation for the Navy's Bureau of Ships; the first prototype was launched 27 December 1955.[1] NASA used Asp as an upper stage in the Nike-Asp briefly: it was test flown several times in 1960, but a need for the vehicle did not develop.[2]

A scientist makes final adjustments to the Nike-Asp payload before launch from Wallops Flight Center.

Black Brant VC

BLACK BRANT. The Black Brant series of sounding rockets was developed by Bristol Aerospace Ltd. of Canada with the Canadian government. The first rocket was launched in 1939. By the end of 1974 close to 300 Black Brants had been launched and vehicles were in inventories of research agencies in Canada, Europe, and the United States, including the U.S. Navy, U.S. Air Force, and NASA. [1]

The Canadian Armament Research and Development Establishment (CARDE) selected the name "Black Brant" for the research rocket, taking the name of a small, dark, fast-flying goose common to the northwest coast and Arctic regions of Canada. The Canadian government kept the name with the addition of numbers (I through VI by 1974) for different members of the series—rather than giving a code name to each version—to emphasize that they were sounding rockets rather than weapons. [2]

NASA took Black Brants into its sounding rocket inventory in 1970 and was using the Black Brant IVA and VC in 1974. The Black Brant IVA used a modified upper stage and a more powerful engine than previous models, to boost it to 900 kilometers. The Black Brant V series consisted of three 43-centimeter-diameter sounding rockets with all components interchangeable.

The Black Brant VA (or "BBVA") used stabilizer components with the BBII's engine and carried 136-kilogram payloads to 160 kilometers, to fill a need for that altitude range. The BBVB, using an engine giving rocket performance double that of the BBII, was designed to meet requirements for scientific investigations above 320-kilometer altitude.

The Black Brant VC was used by NASA to support the 1973–1974 Skylab Orbital Workshop missions by evaluating and calibrating Workshop instruments. The three-fin solid-fueled Black Brant VB was converted to a four-fin model suitable for launching from White Sands Missile Range and permitting recovery of the rocket payloads. The changes decreased performance somewhat but increased stability and allowed greater variations in payload length and weight on the VC. NASA launched the Black Brant VC on two flights during each of the three manned missions to the Skylab Workshop. [3]

The performance range of NASA's Black Brant sounding rockets (with an 85° launch angle) in 1974 was: [4]

Model	Gross Payload Weight (kg)	Altitude (km)
Black Brant IVA	40	900
	100	530
Black Brant VC	200	305

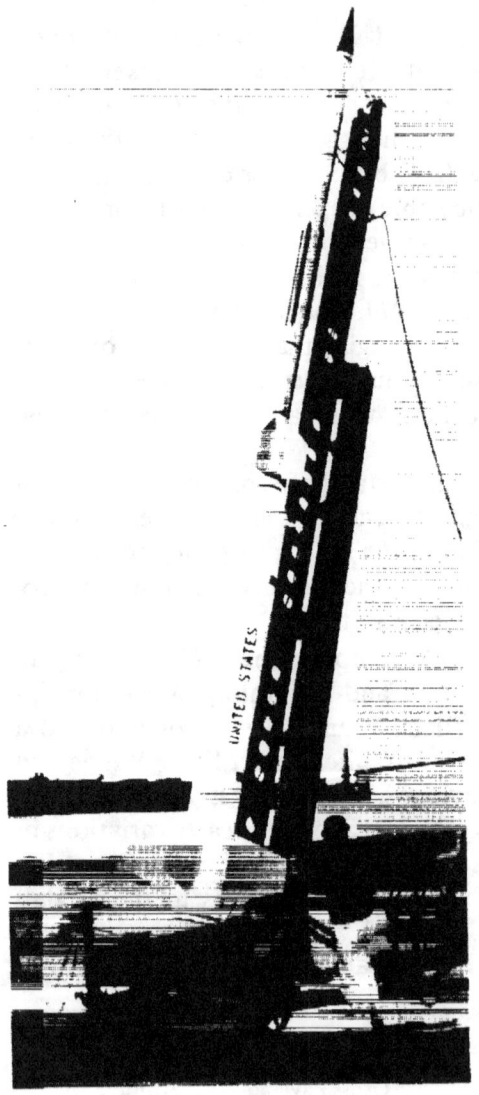

Nike-Cajun in launch position.

The new Hawk launched into flight.

SOUNDING ROCKETS

CAJUN. The Cajun solid-propellant rocket stage was designed and developed under the Pilotless Aircraft Research Division of the National Advisory Committee for Aeronautics' Langley Laboratory (later NASA's Langley Research Center). The project's manager, Joseph G. Thibodaux, Jr., formerly of Louisiana, suggested the new motor be named "Cajun" because of the term's Louisiana associations. It was the name of persons in that region reputed to be of mixed Acadian French and Indian or Negro blood. Allen E. Williams, Director of Engineering in Thiokol Chemical Corporation's Elkton (Md.) Division, agreed to the name, and later the Elkton Division decided to continue giving its rocket motors Indian-related names.[1]

Design of the Cajun motor was based on the Deacon motor, begun during World War II by Allegany Ballistics Laboratory for the National Defense Research Council. NACA purchased Deacon propellant grains from Allegany to propel its aerodynamic research models. Deacon was used with the Nike first stage. In 1956 Langley contracted with Thiokol to develop the improved Deacon, named "Cajun."[2]

The Nike-Cajun, lifting 35-kilogram instrumented payloads to a 160-kilometer altitude, was one of NASA's most frequently used sounding rockets.[3]

HAWK. NASA was developing a low-cost sounding rocket in 1974–1975 using surplus motors from the Army's Hawk antiaircraft missiles. The research rocket inherited the Army's name, an acronym for "Homing All the Way Killer," although the new uses would be far removed from the purposes of the weapon system.[1]

To be flown as a single-stage Hawk or in two-stage combination as the Nike-Hawk, for a variety of research projects, the 35.6-centimeter-diameter rocket would provide a large volume for payloads. Both stages of the Nike-Hawk would use surplus Army equipment (see also Nike). Development testing was proceeding under Wallops Flight Center management. By December 1974, two flight tests of the single-stage Hawk sounding rocket had been launched, the first one lifting off successfully 29 May 1974. The first flight test of the Nike-Hawk was planned for mid-1975.[2]

The single-stage Hawk could carry a 45-kilogram payload to an 80-kilometer altitude or 90 kilograms to 57 kilometers. Engineers were working toward a performance capability of 45 kilograms to 210 kilometers or 90 kilograms to 160 kilometers for the Nike-Hawk.[3]

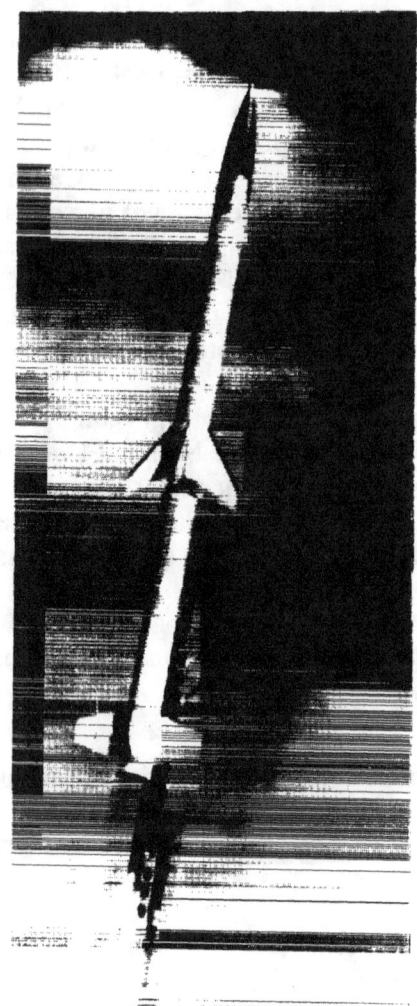

The winged goddess Nike (courtesy of the Library of Congress).

Nike-Malemute lifts off.

Nike-Tomahawk poised for flight.

SOUNDING ROCKETS

MALEMUTE. The Malemute, a rocket second stage, was developed in 1974 in an interagency program with NASA, Sandia Laboratories, and the Air Force Cambridge Research Laboratories as sponsoring agencies. Designed to be flown with either the Nike or the Terrier first stage, the Malemute began flight tests in 1974. It was named for the Alaskan Eskimo people by the contractor, Thiokol Corporation, in Thiokol's tradition of using Indian-related names (see Cajun).[1]

The Nike-Malemute sounding rocket would be able to lift a 90-kilogram payload to 500 kilometers; the Terrier-Malemute would lift the same payload to 700 kilometers. The new vehicles were intended to replace the Javelin and Black Brant IV rockets in the NASA inventory.[2]

NIKE. The Nike, a solid-propellant first stage, was an adaptation of the Nike antiaircraft missile developed, beginning in 1945, by the Hercules Powder Company for U.S. Army Ordnance.[1] The name "Nike" was taken from ancient Greek mythology: Nike was the winged goddess of victory. In NASA's sounding rocket program, Nike was used with Apache, Cajun, Tomahawk, Hawk, or Malemute upper stages, as well as with the Aerobee 170, 200, and 350.[2]

TERRIER. The Terrier, a rocket first stage used by NASA, was developed by Hercules Powder Company as the first stage of the Navy's Terrier antiaircraft missile, and NASA inherited the name. NASA used it with the Malemute second stage, as the Terrier-Malemute.[1]

TOMAHAWK. The Tomahawk, a sounding rocket upper stage used with the Nike booster stage, was named by Thiokol Corporation for the Indian weapon, in Thiokol's tradition of giving its motors Indian-related names (see Cajun).[1] The Nike-Tomahawk could lift 27-kilogram instrumented payloads to a 490-kilometer operating altitude or 118 kilograms to 210 kilometers.[2]

VI

NASA INSTALLATIONS

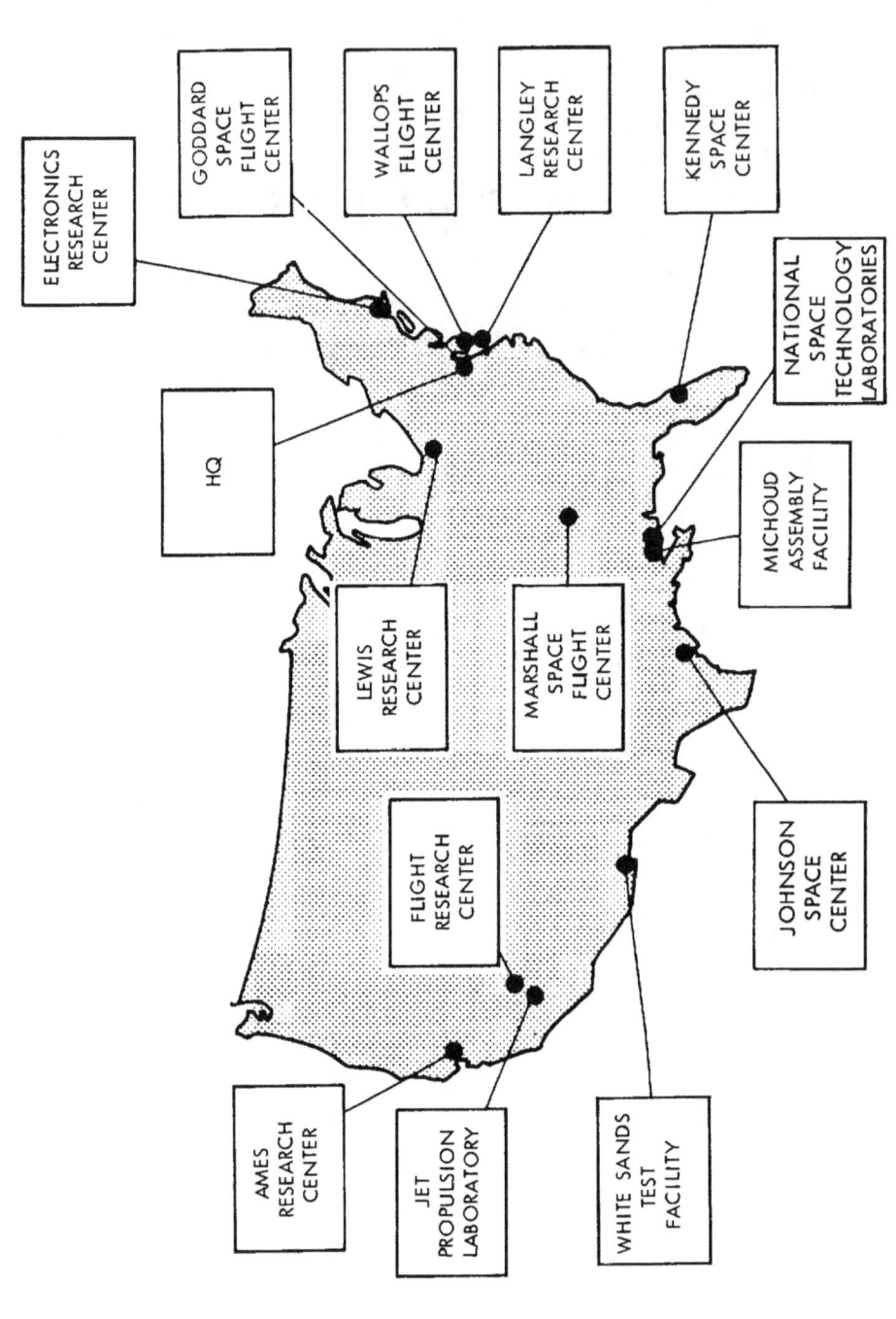

ELECTRONICS RESEARCH CENTER

GODDARD SPACE FLIGHT CENTER

WALLOPS FLIGHT CENTER

LANGLEY RESEARCH CENTER

KENNEDY SPACE CENTER

NATIONAL SPACE TECHNOLOGY LABORATORIES

HQ

MICHOUD ASSEMBLY FACILITY

LEWIS RESEARCH CENTER

MARSHALL SPACE FLIGHT CENTER

FLIGHT RESEARCH CENTER

JOHNSON SPACE CENTER

AMES RESEARCH CENTER

JET PROPULSION LABORATORY

WHITE SANDS TEST FACILITY

NASA INSTALLATIONS

The 11 NASA "field installations" and the contractor-operated Jet Propulsion Laboratory each had a unique history. Many were named for prominent Americans.

Five of the installations were existing facilities of the National Advisory Committee for Aeronautics (NACA), which in October 1958 became the nucleus of the National Aeronautics and Space Administration. These were Ames Research Center, Flight Research Center, Langley Research Center, Lewis Research Center, and Wallops Flight Center.

Three installations—Goddard Space Flight Center, Kennedy Space Center, and Marshall Space Flight Center—and the Jet Propulsion Laboratory began their association with NASA as transfers from the U.S. military space program.

Two installations were created to fill special needs of the U.S. civilian space program. Electronics Research Center joined the research Centers until 1969 and Manned Spacecraft Center, later renamed Johnson Space Center, was added to the manned spaceflight Centers. National Space Technology Laboratories, designated a permanent field installation in 1974, grew out of cooperative activities with other agencies in earth resources and environmental research at an MSFC test facility.

Aerial view of Ames Research Center, above. Dr. Joseph S. Ames at left.

AMES RESEARCH CENTER (ARC). Congress on 9 August 1939 authorized the construction of a second National Advisory Committee for Aeronautics (NACA) laboratory for urgent research in aircraft structures, as World War II began. Ground was broken for the laboratory at Moffett Field, California, 14 September 1939. The NACA facility began operations as the Moffett Field Laboratory in early 1941.[1]

NACA named the facility "Ames Aeronautical Laboratory" in 1944 in honor of Dr. Joseph S. Ames, leading aerodynamicist and former president of Johns Hopkins University, one of the first NACA members in 1915 and serving to 1939. He was NACA Chairman from 1927 to 1939. When Dr. Ames retired as NACA Chairman, he was cited by President Roosevelt for his "inspiring leadership in the development of new research facilities and in the orderly prosecution of comprehensive research programs."[2]

On 1 October 1958, as a facility of the NACA, the laboratory became part of the new National Aeronautics and Space Administration and was renamed "Ames Research Center."[3]

Mission responsibilities of ARC focused on basic and applied research in the physical and life sciences for aeronautics and space flight. The Center managed the Pioneer and Biosatellite space projects, as well as providing scientific experiments for other missions. It contributed to development of experimental tilt-wing and fan-in-wing aircraft and solutions to high-speed atmosphere entry problems.

ELECTRONICS RESEARCH CENTER (ERC). NASA's Electronics Research Center was formally activated 1 September 1964 in Cambridge,

Model of Electronics Research Center.

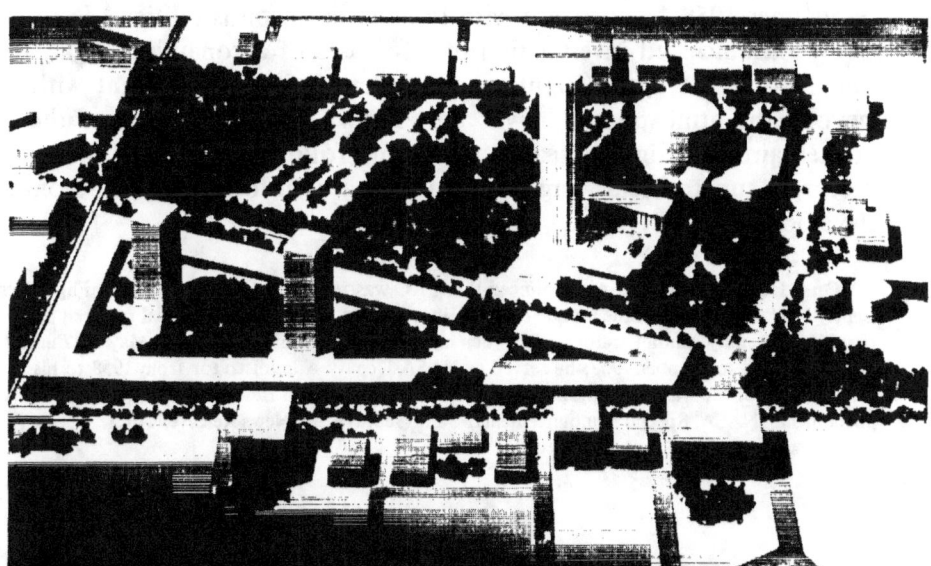

Massachusetts. ERC absorbed the NASA North Eastern Office, established 14 August 1962 to administer contracts and act as liaison with industry in northeastern states.[1] The name "Electronics Research Center" reflected the installation's mission responsibility. As the focal point of nationwide research in this field, the Center organized, sponsored, and conducted comprehensive programs of basic and applied research in space and aeronautical electronics.

On 29 December 1969, NASA announced its decision to close ERC because of budget reductions. The facility was transferred to the Department of Transportation for use in research and development efforts and was renamed the Transportation Development Center.[2]

FLIGHT RESEARCH CENTER (FRC). On 30 September 1946, 13 engineers, instrument technicians, and technical observers were sent on temporary duty from the National Advisory Committee for Aeronautics' Langley Laboratory to assist in the rocket-powered X–1 flight-research program at the Air Force's Muroc, California, test facility. Called the "NACA Muroc Flight Test Unit," this group was the beginning of what was to become the Flight Research Center.[1] In 1949 NACA redesignated the unit—which in 1947 had been permanently assigned at Muroc—the "High Speed Flight Research Station." Muroc Air Force Base itself became Edwards Air Force Base after February 1950. In 1954 the NACA unit moved into new, permanent facilities on 175 acres leased from the Air Force at Edwards and its name was changed to "High Speed Flight Station."[2]

When the National Aeronautics and Space Administration was formed 1 October 1958, the High Speed Flight Station—as a facility of the NACA—became part of NASA. NASA renamed it "Flight Research Center" 27 September 1959,* consistent with its mission responsibilities.[3] Research at the Center included investigation of all phases of aeronautical flight, reentry and landing for space flight, and problems of manned flight within and beyond the atmosphere. It was best known for its conduct of the X–15 rocket aircraft flight research program, followed by X–24 lifting-body research, supercritical wing tests, and research into other new aeronautical development.

---

* On 8 January 1976, NASA announced that FRC was renamed "Hugh L. Dryden Flight Research Center" in "recognition of the unique contributions" of Dr. Dryden, aeronautical research pioneer and first NASA Deputy Administrator. Dr. Dryden had been Director of NACA from May 1947 until NASA was established in October 1958; he served as NASA Deputy Administrator from 1958 to his death in December 1965. He was internationally recognized for his work in fluid mechanics and boundary layer phenomena. (NASA, Special Announcement, 8 Jan. 1956; NASA, News Release 76–7.)

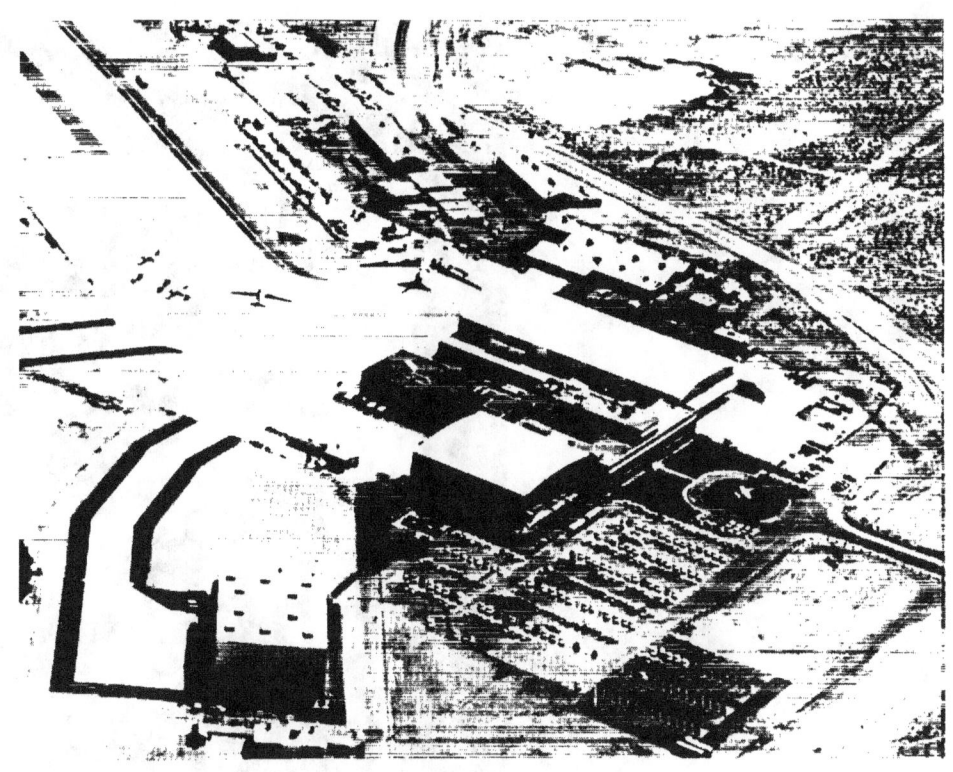

Aerial view of Flight Research Center, above, and a modified X–15 experimental aircraft with external fuel tanks.

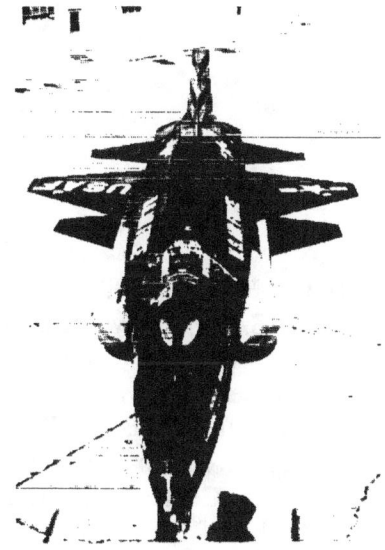

Dr. Robert H. Goddard at work in his shop at Roswell, New Mexico, October 1935 (courtesy of Mrs. Robert H. Goddard).

GODDARD SPACE FLIGHT CENTER (GSFC). In August 1958, before NASA officially opened for business, Congress authorized construction of a NASA "space projects center" in the "vicinity of Washington, D.C." [1] The site selected was in Maryland on land then part of the Department of Agriculture's Beltsville Agricultural Research Center. On 15 January 1959, NASA designated four divisions of NASA Headquarters the "Beltsville Space Center." Project Vanguard personnel, transferred by Executive Order of the President from the Naval Research Laboratory to NASA in December 1958, formed the nucleus of three of the four divisions and hence of the Center. [2] On 1 May 1959, NASA renamed the facility "Goddard Space

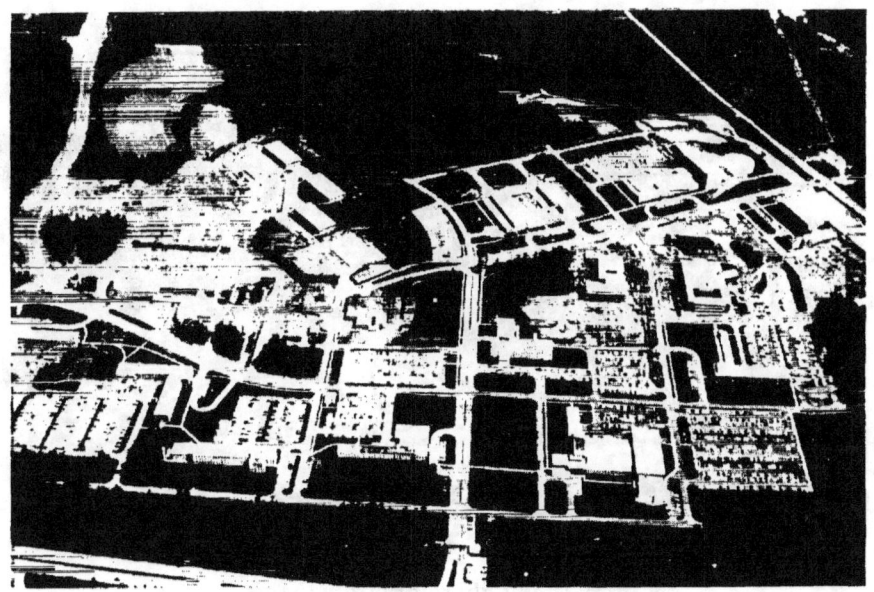

Aerial view of Goddard Flight Research Center.

Flight Center" in honor of the father of modern rocketry, Dr. Robert H. Goddard (1882–1945).[3] Rocket theorist as well as practical inventor, Dr. Goddard's list of "firsts" in rocketry included the first launch of a liquid-propellant rocket March 1926.[4]

Goddard Space Flight Center was responsible for unmanned spacecraft and sounding rocket experiments in basic and applied research; it operated the worldwide Space Tracking and Data Acquisition Network (STADAN), which later became Spaceflight Tracking and Data Network (STDN); and it managed development and launch of the Thor-Delta launch vehicle.

GODDARD INSTITUTE FOR SPACE STUDIES (GISS). A center for theoretical research was established in 1961 as the New York office of the Theoretical Division of Goddard Space Flight Center. In July 1962 it was separated organizationally from the Theoretical Division and renamed "Goddard Institute for Space Studies."[5] It worked closely with academic scientists in the New York area.

JET PROPULSION LABORATORY (JPL). Students at the Guggenheim Aeronautical Laboratory of the California Institute of Technology (GALCIT), directed by Dr. Theodore von Kármán, in 1936 began design

Jet Propulsion Laboratory from the air.

and experimental work with liquid-propellant rocket engines.[1] During World War II the GALCIT Rocket Research Project developed solid- and liquid-propellant units to assist the takeoff of heavily loaded aircraft and began work on high-altitude rockets. Reorganized in November 1944 under the name "Jet Propulsion Laboratory," the facility continued postwar research and development on tactical guided missiles, aerodynamics, and broad supporting technology for U.S. Army Ordnance.[2]

JPL participated with the Army Ballistic Missile Agency in the development and operation of the first U.S. satellite, *Explorer 1*, the succeeding Explorer missions, and the *Pioneer 3* and *4* lunar probes. On 3 December 1958, shortly after NASA came into existence, the functions and facilities of JPL were transferred from the U.S. Army to NASA.[3] Operating in Government-owned facilities, JPL remained a laboratory of Caltech under contract to NASA.* It has managed projects in NASA's unmanned lunar

---

*Public Law 92–520, signed 21 October 1972, carried a rider changing the name of JPL to "H. Allen Smith Jet Propulsion Laboratory" effective 4 January 1973. The House Committee on Public Works had amended a bill for construction of a civic center as a memorial to President Dwight D. Eisenhower, adding a proposal to honor a number of members of Congress by renaming public works buildings after them, including Rep. H. Allen Smith (R–Calif.). Retiring Rep. Smith said he would request legislation to repeal the JPL name change, to avoid confusion, and Public Law 93–215, signed 28 December 1973, included an amendment restoring JPL's name. Jet Propulsion Laboratory never used the H. Allen Smith name. (U.S. Congress, House of Representatives, *Dwight D. Eisenhower Memorial Bicentennial Civic*

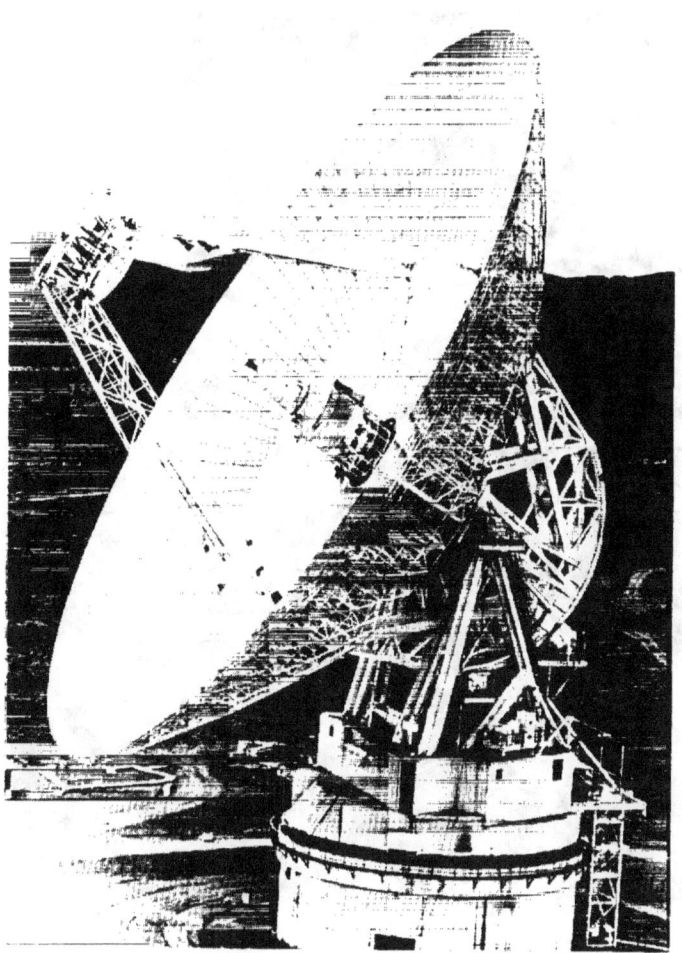

The 64-meter Goldstone antenna of the Deep Space Network.

and planetary exploration program such as Ranger, Surveyor, and the Mariner series, conducted supporting research, and founded and operated the worldwide Deep Space Network (DSN) for communication with lunar and planetary spacecraft.

Center Act, House Rpt. 92–1410, 19 Sept. 1972 [Washington: 1972], pp. 5, 9; U.S. Public Law 92–520, 86 Stat. 1022, 21 Oct. 1972, p. 4; *Wall Street Journal,* 24 Nov. 1972; Pasadena, Calif., *Star-News,* 18 Dec. 1972; Gerald J. Mossinghoff, Deputy Assistant Administrator for Legislative Affairs, NASA, memorandum for the record, 15 Jan. 1973; *Congressional Record—House,* 8 Feb. 1973, p. H932, and 3 Dec. 1973, pp. H10468–H10469; General Services Administration, National Archives and Records Service, Office of the Federal Register, *Weekly Compilation of Presidential Documents,* 10, No. 1 [7 Jan. 1974], p. 15.)

Johnson Space Center seen from the air in the top photo, with the Mission Operations Control Room on the third day of the Apollo 8 lunar orbit mission. The television monitor shows the earth telecast from 283 000 kilometers away. President Lyndon B. Johnson at left.

146

JOHNSON SPACE CENTER (JSC). On 3 January 1961 NASA's Space Task Group—an autonomous subdivision of Goddard Space Flight Center that managed Project Mercury and was housed at Langley Research Center—was made an independent NASA field installation. Following congressional endorsement of President Kennedy's decision to accelerate the U.S. manned spaceflight program, Congress in August 1961 appropriated funds for a new center for manned space flight. On 9 September 1961 NASA announced the "Manned Spacecraft Center" (MSC) would be built at Clear Lake, near Houston, Texas,[1] and on 1 November 1961 Space Task Group personnel were told that "the Space Task Group is officially redesignated the Manned Spacecraft Center."[2]

Known as Manned Spacecraft Center for 11½ years, the Center was responsible for design, development, and testing of manned spacecraft; selection and training of astronauts; and operation of manned space flights—including the Mercury, Gemini, Apollo, and Skylab programs and the U.S.–U.S.S.R. Apollo-Soyuz Test Project. It was lead Center for management of the Space Shuttle program.

Following the 22 January 1973 death of former President Lyndon B. Johnson, leader of support for the U.S. space program from its earliest beginnings, Senator Lloyd M. Bentsen (D–Tex.) proposed that MSC be renamed the "Lyndon B. Johnson Space Center." Senator Robert C. Byrd (D–W. Va.) introduced Senate Joint Resolution 37 on Senator Bentsen's behalf 26 January and House joint resolutions were introduced in the next few days.[3] Support from NASA Headquarters and Manned Spacecraft Center was immediate.[4] The Senate and House acted 6 and 7 February and President Nixon signed the resolution 17 February 1973.[5]

As Senator, Johnson had drafted and helped enact legislation that created NASA. As Vice President he had chaired the National Aeronautics and Space Council during the early years of the space program, when the decision was made to place a man on the moon. As President he continued strong support.[6]

Signing the resolution renaming MSC, President Nixon said, "Lyndon Johnson drew America up closer to the stars, and before he died he saw us reach the moon—the first great platform along the way."[7]

WHITE SANDS TEST FACILITY. In June 1962, Manned Spacecraft Center reached an operating agreement with the U.S. Army's White Sands Missile Range for the establishment of an Apollo propulsion development facility and NASA announced selection of the site.[8] The facility was called "White Sands Operations." NASA renamed the facility the "White Sands Test

A Saturn V test vehicle is transported from the Kennedy Space Center's Vehicle Assembly Building toward Launch Complex 39.

President John F. Kennedy

Facility" 25 June 1965.[9] White Sands was notable in U.S. rocket history as the site for test-firing the German V–2 rockets after World War II.

KENNEDY SPACE CENTER (KSC). Formally named "John F. Kennedy Space Center, NASA," the installation at Cape Canaveral (named Cape Kennedy 1963–1973) evolved through a series of organizational changes and redesignations.

In 1951 the Experimental Missiles Firing Branch of the Army Ordnance Guided Missile Center in Huntsville, Alabama, was established to supervise test flights of the U.S. Army's Redstone intermediate-range ballistic missile at the Long Range Proving Ground at Cape Canaveral, Florida. In January 1953, when its responsibilities were expanded, the Army facility was renamed "Missile Firing Laboratory." [1]

On 1 July 1960 the Missile Firing Laboratory became part of NASA's Marshall Space Flight Center (MSFC)—the nucleus of which was the Laboratory's parent organization at Huntsville—and it was absorbed organizationally into MSFC's Launch Operations Directorate.[2] The other basic element of the Launch Operations Directorate was a NASA unit known as "AMROO" (Atlantic Missile Range Operations Office). AMROO had functioned as NASA's liaison organization with the military-operated Atlantic Missile Range (formerly Long Range Proving Ground) at Cape Canaveral. Together, the Missile Firing Laboratory and AMROO formed MSFC's Launch Operations Directorate.[3]

The Launch Operations Directorate was discontinued as a component of MSFC on 7 March 1962 and Launch Operations Center was established as a separate NASA field installation, officially activated 1 July 1962.[4] On 29 November 1963, a week after the death of President John F. Kennedy, President Lyndon B. Johnson renamed the Launch Operations Center the "John F. Kennedy Space Center," saying that President Kennedy had "lighted the imagination of our people when he set the moon as our target and man as the means to reach it" and that the Center was a "symbol of our country's peaceful assault on space." [5]

Adjacent to Cape Canaveral was the 324-square-kilometer Merritt Island. In the autumn of 1961 NASA had selected it for launches in the Apollo manned lunar program.[6] On 17 January 1963 the Launch Operations Center became the executive agent for management and operation of the "Merritt Island Launch Area" (usually called "MILA").[7] Headquarters of Kennedy Space Center moved to new facilities on Merritt Island 26 July 1965, and NASA discontinued the "MILA" designation, calling the entire NASA complex the Kennedy Space Center.[8] The Center was responsible for overall NASA launch operations at the Eastern Test Range (formerly Atlantic

Missile Range), Western Test Range, and KSC itself, including launches of satellites, probes, manned space missions, and the Space Shuttle.

NASA DAYTONA BEACH OPERATIONS. The Daytona Beach facility was established at the General Electric Company in Daytona Beach, Florida, 23 June 1963 as liaison between NASA and GE; it was an integral part of the Launch Operations Center (later Kennedy Space Center). [9]

WESTERN TEST RANGE OPERATIONS DIVISION. The WTR facility originated 27 October 1960, when NASA established the Test Support Office at the Pacific Missile Range (PMR) for liaison between NASA and the military-operated PMR. The Test Support Office came under the jurisdiction of MSFC's Launch Operations Directorate. NASA discontinued the Test Support Office 7 March 1962 and established the Pacific Launch Operations Office at PMR as an independent field installation. [10] On 1 October 1965 the Pacific Launch Operations Office and the Launch Operations Division of Goddard Space Flight Center at the Western Test Range (formerly Pacific Missile Range) were combined to form the Western Test Range Operations Division of KSC. [11]

LANGLEY RESEARCH CENTER (LaRC). Construction of NACA's first field station began at Langley Field near Hampton, Virginia, in 1917. In April 1920, President Wilson concurred with NACA's suggestion that the facility be named "Langley Memorial Aeronautical Laboratory" in honor of Dr. Samuel P. Langley (1834–1906). [1] Dr. Langley was the third Secretary of the Smithsonian Institution, "inventor, brilliantly lucid writer and lecturer on science, original investigator in astrophysics and especially of the physics of the sun, pioneer in aerodynamics." He was "all this and more." His persistent investigation of mechanical flight led to successful flights by his steam-powered, heavier-than-air "aerodromes" in 1896; on 6 May his model made two flights, each close to 1 kilometer long, and on 28 November his aerodrome achieved a flight of more than 1.2 kilometers. [2]

The facility was dedicated 11 July 1920, marking "the real beginning of NACA's own program of aeronautical research, conducted by its own staff in its own facilities." [3] It was the only NACA laboratory until 1940. On 1 October 1958 the laboratory, as a NACA facility, became a component of the National Aeronautics and Space Administration and was renamed "Langley Research Center." [4]

The Center conducted basic research in a variety of fields for aeronautical and space flight and had management responsibility for the Lunar Orbiter and Viking projects and the Scout launch vehicles. The supercritical wing, an improved airfoil, was developed at Langley.

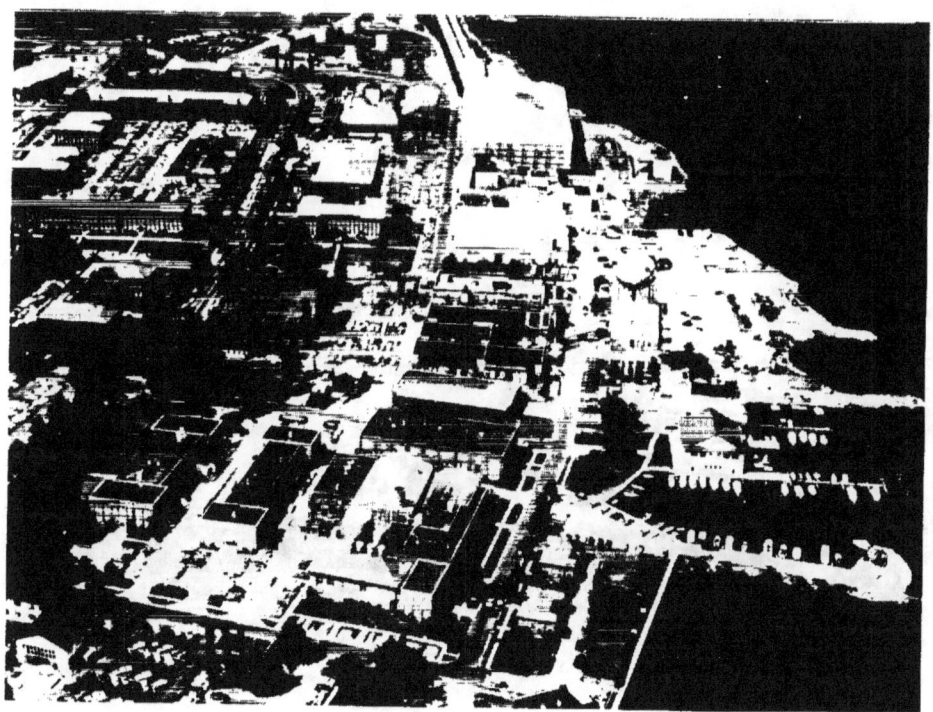

Facilities of Langley Research Center

Samuel P. Langley (courtesy Smithsonian Institution)

Lewis Research Center

LEWIS RESEARCH CENTER (LeRC). Congress authorized a flight-propulsion laboratory for NACA 26 June 1940, and in 1942 the new laboratory began operations adjacent to the Cleveland, Ohio, Municipal Airport.[1] It was known as the "Aircraft Engine Research Laboratory." [2] On 28 September 1948 NACA renamed it "Lewis Flight Propulsion Laboratory" in honor of Dr. George W. Lewis (1882–1948). Dr. Lewis not only was a leading aeronautical engineer, whose work in flight research has been termed "epochal contributions to aeronautics," [3] but also made his mark as an administrator, serving as NACA's Director of Aeronautical Research from 1919 to 1947. He was responsible for the planning and building of the new flight-propulsion laboratory which was later to bear his name.[4]

Upon the formation of the National Aeronautics and Space Administration 1 October 1958, the facility became "Lewis Research Center." [5] The Center's research and development responsibilities concentrated chiefly on advanced propulsion and space power systems. It had management responsibilities for the Agena and Centaur launch vehicle stages.

Dr. George W. Lewis

PLUM BROOK STATION. On Lake Erie near Sandusky, Ohio, Lewis Research Center's Plum Brook Station was a test facility for aerospace propulsion research and development. The site, formerly a U.S. Army Ordnance plant, was acquired from the Army through a gradual process beginning in 1956 and completed in 1963. The name "Plum Brook Station" derived from the Army's name of the former ordnance facility, "Plum Brook Ordnance Works," after a small stream running through the site.[6] It had a nuclear research reactor and a wide range of propulsion test facilities.

Nuclear propulsion program cutbacks to adjust to NASA budget reductions in 1973 brought a decision to phase down most of the Plum Brook facilities. The Space Power Facility—one of the world's largest space environment simulation chambers, equipped with a solar simulation system, instrumentation, and data-acquisition facilities—was kept in operation for use by other Government agencies. The Air Force, Navy, National Oceanic and Atmospheric Administration, and Energy Research and Development Administration indicated possible interest in using the facilities.

By the end of June 1974, agencies already using the facilities at the station included the Army, Ohio National Guard, and Department of the Interior. Instrumentation for a large experimental wind generator was being installed in the cooperative NASA and National Science Foundation program to study full-scale wind-driven energy devices.[7]

153

Three-building headquarters complex of Marshall Space Flight Center.

MARSHALL SPACE FLIGHT CENTER (MSFC). In April 1950 the U.S. Army established its team of rocket specialists headed by Dr. Wernher von Braun as the Ordnance Guided Missile Center at Redstone Arsenal, Huntsville, Alabama. This Center was the origin of what eventually became the George C. Marshall Space Flight Center (MSFC). On 1 February 1956 the Army Ballistic Missile Agency (ABMA) was formed at Redstone Arsenal. ABMA was a merger and expansion of existing agencies there; its team of scientists formed the nucleus of the Development Operations Division.[1]

NASA came into existence on 1 October 1958. Early in 1960 President Eisenhower submitted a request to Congress for the transfer of ABMA's space missions to NASA, including certain facilities and personnel, chiefly the Development Operations Division. The transfer became effective 14 March 1960 and NASA set up its "Huntsville Facility" in preparation for formal establishment of the field center later that year.[2] The next day, 15 March, President Eisenhower proclaimed the NASA facility would be called "George C. Marshall Space Flight Center."[3] The name honored George C. Marshall, General of the Army, who was Chief of Staff during World War

II, Secretary of State 1948–1949, and author of the Marshall Plan. General Marshall was the only professional soldier to receive the Nobel Peace Prize, awarded to him in 1954.

MSFC officially began operation with the formal mass transfer of personnel and facilities from ABMA 1 July 1960.[4] The Center's primary mission responsibility was development of the Saturn family of launch vehicles, used in the Apollo manned lunar-landing program, in the Skylab experimental space station program, and in the U.S.–U.S.S.R. Apollo-Soyuz Test Project. MSFC also held responsibility for development of the Skylab Orbital Workshop and Apollo Telescope Mount, as well as integration of the Skylab cluster of components. It was responsible for three major elements of the Space Shuttle: the solid-fueled rocket booster, the Space Shuttle main engine, and the external tank.

MICHOUD ASSEMBLY FACILITY. On 7 September 1961 NASA selected the Government-owned, then-unused Michoud Ordnance Plant at Michoud, Louisiana, as the site for industrial production of Saturn launch vehicle stages under the overall direction of Marshall Space Flight Center. NASA called the site "Michoud Operations." [5] On 1 July 1965 Michoud Operations was redesignated "Michoud Assembly Facility" to "better reflect the mission" of the facility.[6]

Following construction of the first stages of the Saturn IB and Saturn V launch vehicles for Apollo, Skylab, and ASTP missions, Michoud was

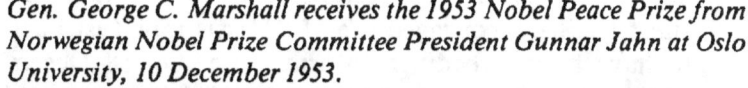

Gen. George C. Marshall receives the 1953 Nobel Peace Prize from Norwegian Nobel Prize Committee President Gunnar Jahn at Oslo University, 10 December 1953.

Final touches are added to engines of the first flight Saturn V's S–IC stage, assembled at Michoud Assembly Facility.

selected in 1972 as the site for the manufacture and final assembly of the Space Shuttle's external propellant tanks.

MISSISSIPPI TEST FACILITY (MTF). NASA announced 25 October 1961 it had selected southwestern Mississippi as the site for a large booster (Saturn) test facility under the direction of MSFC. Pending official naming of the site,* NASA encouraged use of "Mississippi Test Facility," which seemed to have been already in informal use. On 18 December 1961 the name "Mississippi Test Operations" was officially adopted, but the site was still widely called "Mississippi Test Facility," particularly by Headquarters and MSFC offices concerned in the installation's development.[7] On 1 July 1965 MSFC announced the official redesignation, "Mississippi Test Facility." [8] The change was said to "reflect the mission of the facility" better.**

MTF test stands were put into standby status 9 November 1970, after more than four years and the test-firing of 13 S–IC first stages and 15 S–II second stages of the Saturn V.[9] With the close of Saturn production and the approaching end of the Apollo program, NASA had established an Earth

---

*For about a month there was no standardized designation and "Pearl River Test Site" was often used. See NASA, Circular 188.

**Perhaps a more accurate reason would be that this was the name most widely used; the "Mississippi Test Operations" never had stuck.

Resources Laboratory at MTF in September 1970, stressing applications of remote-sensing data from aircraft and satellites. A number of other Government agencies, at NASA invitation, moved research activities related to resources and the environment to MTF to take advantage of its facilities.[10] And on 1 March 1971 NASA announced that MTF had also been selected for sea-level testing of the Space Shuttle's main engine.[11]

On 14 June 1974 Mississippi Test Facility was renamed "National Space Technology Laboratories" and became a permanent NASA field installation reporting directly to NASA Headquarters, "because of the growing importance of the activities at NSTL . . . and of the agencies taking advantage of NSTL capabilities."[12] (See National Space Technological Laboratories.)

NATIONAL SPACE TECHNOLOGY LABORATORIES (NSTL). Established as an independent NASA field installation 14 June 1974,[1] the National Space Technology Laboratories' varied activities had their beginnings in the Mississippi Test Facility (MTF), formed in 1961 as part of Marshall Space Flight Center to test Saturn launch vehicle stages (see Mississippi Test Facility). The facility at Bay St. Louis in Hancock County, Mississippi, tested the

Test stand A–1 at National Space Technology Laboratories was modified for testing the Space Shuttle main engine.

Saturn V first and second stages throughout that program, qualifying them for the Apollo and Skylab missions. With the shift of emphasis in the national space program from manned exploration to practical applications after the successful *Apollo 11* landing on the moon 20 July 1969, and as the last lunar exploration missions were made 1970-1972, consideration was given to other uses for the MTF plant. Increasing awareness of the importance of the earth's natural resources and environment in those years—and sharpening focus on energy shortages in 1973—suggested that technical facilities available at MTF might be put to use in meeting some of these problems.

MTF test stands were put on standby as of 9 November 1970.[2] NASA had already established an Earth Resources Laboratory at the installation in September 1970 and had invited other Government agencies to use facilities on the 570-square-kilometer site for research.[3] In 1971 MTF was also selected for development testing of the main engine for NASA's Space Shuttle, designed as a reusable, economical space transportation system for the 1980s.[4]

By June 1974 a number of other agencies had established one or more activities at NSTL: the Department of Commerce, Department of the Interior, Department of Transportation, Department of the Army, the U.S. Environmental Protection Agency, the State of Mississippi, and some other state and university elements from Mississippi and Louisiana. Research and technical activities were primarily related to earth resources and the environment. NASA's Earth Resources Laboratory complemented programs at Goddard Space Flight and Johnson Space Centers and emphasized applications of data gathered by remote sensing from aircraft and satellites.[5]

NASA Administrator James C. Fletcher announced the new name and status of National Space Technology Laboratories 14 June 1974, saying that the success of the experiment in collocating "mutually supporting activities" had led him to decide that "NSTL will have a permanent role in NASA's space applications and technology programs." NASA would encourage location at NSTL of other Government activities that could use and contribute to the capabilities there.[6]

WALLOPS FLIGHT CENTER (WFC). The National Advisory Committee for Aeronautics (NACA) established a test-launching facility for its Langley Laboratory on Wallops Island, Virginia, 7 May 1945. A unit of Langley, it was named the "Auxiliary Flight Research Station." On 10 June 1946, the unit became a division of Langley's Research Department and was named "Pilotless Aircraft Research Division (PARD)." The phrase "pilotless air-

Aerial view of Wallops Flight Center main base.

craft" was then used by the Navy's Bureau of Aeronautics and the Army Air Forces to denote all guided missiles. PARD was formally organized 11 August 1946, with four sections; the Wallops facility was placed under PARD's Operations Section and named "Pilotless Aircraft Research Station." Its employees called the station simply "Wallops." [1]

When the National Aeronautics and Space Administration absorbed NACA in 1958, NASA continued the name long in popular use, "Wallops Station" (WS). WS first was carried on organization charts as coming under the proposed Space Flight Research Center, but on 1 May 1959 the station became an independent field installation. [2]

The island—and hence the installation (which in July 1959 acquired additional property on the mainland, known as Wallops Main Base [3])—was named for the 17th-century surveyor John Wallop, who began patenting land on Virginia's eastern shore in the 1660s. In 1672 he received a Crown Patent of the 13-square-kilometer island from King Charles II, and in his will John Wallop referred to "my island formerly called Keeckotank." (It was also known as Accocomoson or Occocomoson Island.) It has borne the name "Wallops Island" for more than 260 years. [4]

Effective 26 April 1974, Wallops Station was renamed "Wallops Flight Center" as "more descriptive of the mission and operations" of the installation. The only rocket flight-test range owned and operated by NASA, Wallops launched Scout boosters and sounding rocket experiments with instrumentation developed by scientists and engineers throughout the United States and the world. By the spring of 1974 more than 8000 launchings had taken place from WFC, including the orbiting of 17 satellites. Work also included advanced aeronautical research and participation in the Chesapeake Bay Ecological Test Program, with remote sensing of the area by aircraft and satellite.[5]

APPENDIX A

SELECTED LIST OF ABBREVIATIONS, ACRONYMS, AND TERMS

This highly selective list includes designations of flight hardware subsystems and components as well as designations of nonflight hardware used in NASA's aeronautics and space research. Many of these terms are components of the projects listed in Parts I–V of the text; others fall outside the project names. Not listed are subsystems of most launch vehicles.

ACS	Attitude control system
AD	Air density satellite (Explorer)
AE	Atmosphere Explorer satellite
AEROSAT	Joint FAA–ESRO Aeronautical Satellite
AIMP	Anchored Interplanetary Monitoring Platform (Explorer)
ALFMED	Apollo light-flash moving-emulsion detection
ALSEP	Apollo lunar surface experiments package
AM	Airlock module (Skylab spacecraft component)
AMPS	Atmospheric, Magnetospheric Plasma-in-Space laboratory (Space Shuttle payload)
AMU	Astronaut maneuvering unit
Anik	Canadian Telesat domestic telecommunications satellite
ANNA	Army-Navy-NASA geodetic satellite
APS	Ascent propulsion system (Apollo LM component); also auxiliary propulsion system
APT	Automatic picture transmission
ASTP	Apollo-Soyuz Test Project
ATDA	Augmented target docking adapter
ATM	Apollo Telescope Mount (Skylab spacecraft component)
AVCS	Advanced vidicon camera system
BBIVA, BBVC	Black Brant IVA, Black Brant VC sounding rockets
BE	Beacon Explorer satellite
BIC	Barium-ion-cloud experiment
BIOCORE	*Apollo 17* medical experiment (using mice)
BIOS	Biological Investigation of Space (suborbital flight experiment); sometimes also used as short name for Biosatellite
CAS	Cooperative Applications Satellite (CAS–A was *Eole*)
CEPE	Cylindrical Electrostatic Probe Experiment (orbital experiment attached to Delta second stage)

CM	Command module (Apollo spacecraft)
Cos	ESRO cosmic ray satellite
CRISP	Cosmic Ray Ionization Program
CSM	Command and service module (Apollo spacecraft, combination of CM and SM)
CTS	Communications Technology Satellite
DAD	Dual Air Density Explorer satellite
DFBW	Digital fly-by-wire program
DME	Direct Measurement Explorer satellite
DPS	Descent propulsion system (Apollo LM component)
DSN	Deep Space Network
EASEP	Early Apollo scientific experiments package (*Apollo 11*)
ECS '	Environmental control system
EGO	Eccentric (orbiting) Geophysical Observatory
EOPAP	Earth and ocean physics applications program
ERAP	Earth resources aircraft program
EREP	Earth resources experiment package (Skylab experiment)
EVA	Extravehicular activity
EXAMETNET	Experimental InterAmerican Meteorological Rocket Network
EXOSAT	ESRO high-energy astronomy satellite, for x-ray astronomy
FIRE	Flight Investigation of the Reentry Environment (reentry heating project)
GATV	Gemini Agena Target Vehicle
GEOS	Geodetic Satellite (Explorer); also its successor, Geodynamic Experimental Ocean Satellite, and ESRO's Geostationary Scientific Satellite
GOES	Geostationary Operational Environmental Satellite
GRS	German Research Satellite (GRS–1 was *Azur*; GRS–2–A, *Aeros*)
GTS	Geostationary Technology Satellite
HAPPE	High Altitude Particle Program Experiment
Hawkeye	Scientific satellite (Explorer; follow-on to Injun series)
HCMM	Heat Capacity Mapping Mission Explorer
HET	Health/education telecommunications experiment (on *ATS 6*)
HL–10	Lifting-body research vehicle
IE	Ionosphere Explorer
IME	Interplanetary meteoroid experiment
IMP	Interplanetary Monitoring Platform (Explorer)
Injun	Scientific satellite (Explorer)
IRLS	Interrogation, recording, and location system
ISEE	International Sun-Earth Explorer
IU	Instrument unit
IUE	International Ultraviolet Explorer
IUS	Interim upper stage for Space Shuttle (called orbit-to-orbit stage, or OOS, by Air Force; interim version of Space Tug)
Kiwi	Ground-test reactor for nuclear propulsion research
LC	Launch complex

ABBREVIATIONS, ACRONYMS

LDEF	Long-Duration Exposure Facility (satellite)
LES	Launch escape system .
LEM	Lunar excursion module (renamed LM)
LLRV	Lunar landing research vehicle
LLTV	Lunar landing training vehicle
LM	Lunar module (Apollo lunar-landing spacecraft)
LOI	Lunar orbit insertion
LOR	Lunar orbit rendezvous (Apollo lunar-landing mode)
LRV	Lunar roving vehicle (Rover)
LST	Large Space Telescope (satellite)
Luster, Project	Sounding Rocket experiment to capture interplanetary particles
M2–F2	Lifting-body research vehicle
MAROTS	ESRO Maritime Orbital Test Satellite
MCC	Midcourse correction; also Mission Control Center
MDA	Multiple docking adapter (Skylab spacecraft component)
MDS	Meteoroid Detection Satellite (Explorer)
MESA	Modularized equipment stowage assembly (Apollo LM component)
METEOSAT	ESRO geostationary meteorological satellite
MET; METS	Mobile equipment transporter; modularized equipment transport system (for Apollo lunar landing missions)
MSFN	Manned Space Flight Network
MTS	Meteoroid Technology Satellite
NERV	Nuclear Emulsion Recovery Vehicle (high-altitude radiation experiment)
NERVA	Nuclear engine for rocket vehicle application
NOAA	National Oceanic and Atmospheric Administration meteorological satellite (successor to ESSA satellites)
NOMSS	National Operational Meteorological Satellite System
OOS	Orbit-to-orbit stage (see IUS)
OSCAR	Orbiting Satellite Carrying Amateur Radio
OTS	ESRO Orbital Test Satellite
OWL	Rice University scientific satellite (Explorer)
OWS	Orbital Workshop (Skylab space station)
PAET	Planetary Atmosphere Experiment Test (suborbital flight experiment)
Phoebus	Ground-test reactor for nuclear propulsion
PLACE	Position location and communications experiment (on $ATS\ 6$)
PLSS	Portable life support system
POGO	Polar Orbiting Geophysical Observatory
Prometheus	Sounding rocket for lightning research
PS	Payload shroud
QUESTOL	Quiet, experimental, short takeoff and landing aircraft
RAE	Radio Astronomy Explorer
RAM	Research and Applications Module (Spacelab forerunner)
RAM, Project	Radio attenuation measurement (reentry communications blackout research)
RAM C–I, etc.	Suborbital spacecraft in Project RAM

RBV	Return-beam-vidicon camera
RCS	Reaction control system
RM	Radiation/Meteoroid satellite
RMU	Remote maneuvering unit
Rover, Project	NASA–AEC research and development program of nuclear reactor propulsion for rockets
RPRV	Remotely piloted research vehicle
RTG	Radioisotope thermoelectric generator (spacecraft power system)
S–I	Saturn I booster first stage
S–IB	Saturn IB booster first stage
S–IC	Saturn V booster first stage
S–II	Saturn V second stage
S–IV	Saturn I second stage
S–IVB	Saturn IB second stage; Saturn V third stage
SAS	Small Astronomy Satellite (Explorer); also solar array system (Skylab Orbital Workshop component)
Scanner, Project	Horizon definition experiment (sounding rocket)
Scramjet	Supersonic combustion ramjet engine
SCS	Stabilization and control system
SE	Solar Explorer satellite
SERT	Space Electric Rocket Test (SERT 1 was suborbital; *SERT 2* was orbited)
SEVA	Surface extravehicular activity (lunar exploration); also standup extravehicular activity
SHAPE	Supersonic High Altitude Parachute Experiment
SIM	Scientific instrument module (Apollo SM component)
SITE	Satellite instructional television experiment (on *ATS 6*)
SLA	Spacecraft–lunar module adapter; also spacecraft launch vehicle adapter
SM	Service module (Apollo spacecraft component)
SNAP	Systems for nuclear auxiliary power (nuclear-electric spacecraft power supply)
SPAN	Solar Particle Alert Network
SPANDAR	Space and range radar
SPED	Supersonic Planetary Entry Decelerator (suborbital flight test)
SPS	Service propulsion system (Apollo SM component)
SRB	Solid (fueled) rocket booster (for Space Shuttle)
SRM	Solid (fueled) rocket motor
SSME	Space Shuttle main engine
SSS	Small Scientific Satellite (Explorer)
SST	Supersonic transport aircraft
STA	Shuttle training aircraft
STADAN	Space Tracking and Data Acquisition Network (see also STDN)
STDN	Spacecraft Tracking and Data Network (formerly STADAN)
STOL	Short takeoff and landing aircraft
STS	Space transportation system
TETR	Test and Training Satellite (MSFN training satellite; see also TTS)
TLI	Translunar injection (insertion into trajectory for the moon)
TOPS	Thermoelectric outer planets spacecraft

ABBREVIATIONS, ACRONYMS

Topsi	Topside sounder satellite
TTS	Test and Training Satellite (MSFN training satellite; see also TETR)
V/STOL	Vertical or short takeoff and landing aircraft
VTOL	Vertical takeoff and landing aircraft
X–15	High-altitude, high-speed, experimental rocket research aircraft
X–24	Lifting-body research vehicle
XB–70	Experimental supersonic aircraft

APPENDIX B
INTERNATIONAL DESIGNATION OF SPACECRAFT

COPY OF THE TEXT OF A NEWS RELEASE ISSUED 31 DECEMBER 1962
BY THE NATIONAL ACADEMY OF SCIENCES–NATIONAL RESEARCH COUNCIL

Beginning on January 1, the international system for designating satellites and space probes for scientific purposes will be changed; as of the new year, Arabic numerals will supplant Greek letters in the satellite designation system.

Prior to January 1, satellites were named in the order of the letters of the Greek alphabet, beginning anew each year: the first satellite launched (Sputnik I) was 1957 Alpha, the first 1958 satellite (Explorer I) was 1958 Alpha, the second (Vanguard I) was 1958 Beta, and so on. The first satellite or space probe in 1963 will be 1963–1, the second will be 1963–2, etc. The numbering will also begin anew each year; for example, the fifth space vehicle of 1964 will be 1964–5.

Usually the launching of a satellite places more than one object in orbit. Sometimes two or more satellites are carried into space where they are separated and ejected into separate orbits. Moreover, the burned-out rocket casing also goes into orbit. The new system provides that the suffix A will identify the main satellite or space probe (i.e., the one carrying the principal scientific payload), and that B, C, etc., as needed, will be used first for any subsidiary scientific payloads in separate orbits, and then for inert components. Thus, under the old system the navigation satellite, Transit II–A, its piggyback companion, Greb, and the spent rocket which injected them into orbit, were called 1960 Eta 1, 1960 Eta 2, and 1960 Eta 3, respectively. If the new scheme had been in effect, they would have been called 1960–7A, 1960–7B, and 1960–7C, respectively.

The new system was agreed upon by all national members of the Committee on Space Research (including both satellite-launching nations) at its meeting in Washington, May 1962. The Committee on Space Research (COSPAR) was established by the International Council of Scientific Unions to facilitate international cooperation in space research. U.S. membership in COSPAR is effectuated through the National Academy of Sciences.

In the United States, the new system will be adopted by the National Aeronautics and Space Administration and the Department of Defense. It will also be used in registering U.S. satellites and space probes with the United Nations.

APPENDIX C

NASA MAJOR LAUNCH RECORD, 1958–1974

This list is a compilation of launches by NASA of (1) payloads that went into orbit or that achieved an altitude of at least 6400 kilometers and (2) major suborbital flight tests or experiments. Included are U.S. launches conducted by the Army Ballistic Missile Agency (ABMA) and the Naval Research Laboratory (NRL) before the establishment of NASA, 1 October 1958. Not listed are launch failures or sounding rocket launches.

Name	International Designation*	Launch Date**	Launch Vehicle***
Explorer 1 (ABMA)	1958 Alpha 1	31 January 1958	Juno I
Vanguard 1 (NRL)	1958 Beta 2	17 March 1958	Vanguard
Explorer 3 (ABMA)	1958 Gamma 1	26 March 1958	Juno I
Explorer 4 (ABMA)	1958 Epsilon 1	26 July 1958	Juno I
Pioneer 1	1958 Eta 1	11 October 1958	Thor-Able I
Pioneer 3	1958 Theta 1	6 December 1958	Juno II
Vanguard 2	1959 Alpha 1	17 February 1959	Vanguard
Pioneer 4	1959 Nu 1	3 March 1959	Juno II

*For simplicity, manned Apollo flights are represented in this list by single designations (lunar orbital and landing missions have separate letter designations in the international system for lunar module and S–IVB stage). Suborbital flights are not assigned international designations.

**Date given is determined by local time at the launch site.

***Thor-Delta launch vehicle configurations are abbreviated as follows: Thor-Delta (Thor-Delta, Thor-improved Delta), TAT–Delta (thrust-augmented Thor-Delta), TAID (thrust-augmented Thor-improved Delta), LTTAT–Delta (long-tank, thrust-augmented Thor-improved Delta), TAT-Agena (thrust-augmented Thor-Agena).

Name	International Designation*	Launch Date**	Launch Vehicle***
Explorer 6	1959 Delta 1	7 August 1959	Thor-Able III
Big Joe	(suborbital)	9 September 1959	Atlas
Vanguard 3	1959 Eta 1	18 September 1959	Vanguard
Little Joe 1	(suborbital)	4 October 1959	Little Joe
Explorer 7	1959 Iota 1	13 October 1959	Juno II
Shotput 1	(suborbital)	28 October 1959	Shotput
Little Joe 2	(suborbital)	4 November 1959	Little Joe
Little Joe 3	(suborbital)	4 December 1959	Little Joe
Shotput 2	(suborbital)	16 January 1960	Shotput
Little Joe 4	(suborbital)	21 January 1960	Little Joe
Shotput 3	(suborbital)	27 February 1960	Shotput
Pioneer 5	1960 Alpha 1	11 March 1960	Thor-Able IV
Shotput 4	(suborbital)	1 April 1960	Shotput
Tiros 1	1960 Beta 2	1 April 1960	Thor-Able
Shotput 5	(suborbital)	31 May 1960	Shotput
Echo 1	1960 Iota 1	12 August 1960	Thor-Delta
Scout 2	(suborbital)	4 October 1960	Scout
Explorer 8	1960 Xi 1	3 November 1960	Juno II
Little Joe 5	(suborbital)	8 November 1960	Little Joe
Tiros 2	1960 Pi 1	23 November 1960	Thor-Delta
Mercury-Redstone 1A	(suborbital)	19 December 1960	Redstone
Mercury-Redstone 2	(suborbital)	31 January 1961	Redstone
Explorer 9	1961 Delta 1	16 February 1961	Scout
Mercury-Atlas 2	(suborbital)	21 February 1961	Atlas
Little Joe 5A	(suborbital)	18 March 1961	Little Joe
Mercury-Redstone BD	(suborbital)	24 March 1961	Redstone
Explorer 10	1961 Kappa 1	25 March 1961	Thor-Delta
Explorer 11	1961 Nu 1	27 April 1961	Juno II
Little Joe 5B	(suborbital)	28 April 1961	Little Joe

NASA MAJOR LAUNCH RECORD

Name	International Designation*	Launch Date**	Launch Vehicle***
Freedom 7 (Mercury-Redstone 3)	(suborbital)	5 May 1961	Redstone
Tiros 3	1961 Rho 1	12 July 1961	Thor-Delta
Liberty Bell 7 (Mercury-Redstone 4)	(suborbital)	21 July 1961	Redstone
Explorer 12	1961 Upsilon 1	15 August 1961	Thor-Delta
Ranger 1	1961 Phi 1	23 August 1961	Atlas-Agena B
Explorer 13	1961 Chi 1	25 August 1961	Scout
Mercury-Atlas 4	1961 Alpha Alpha 1	13 September 1961	Atlas D
Probe A (P–21)	(suborbital)	19 October 1961	Scout
Saturn-Apollo 1 (SA–1)	(suborbital)	27 October 1961	Saturn I
Ranger 2	1961 Alpha Theta 1	18 November 1961	Atlas-Agena B
Mercury-Atlas 5	1961 Alpha Iota 1	29 November 1961	Atlas D
Echo (AVT–1)	(suborbital)	15 January 1962	Thor
Ranger 3	1962 Alpha 1	26 January 1962	Atlas-Agena B
Tiros 4	1962 Beta 1	8 February 1962	Thor-Delta
Friendship 7 (Mercury-Atlas 6)	1962 Gamma 7	20 February 1962	Mercury-Atlas D
Reentry 1	(suborbital)	1 March 1962	Scout
OSO 1	1962 Zeta 1	7 March 1962	Thor-Delta
Probe B (P–21A)	(suborbital)	29 March 1962	Scout
Ranger 4	1962 Mu 1	23 April 1962	Atlas-Agena B
Saturn-Apollo 2 (SA–2)	(suborbital)	25 April 1962	Saturn I
Ariel 1	1962 Omicron 1	26 April 1962	Thor-Delta
Aurora 7 (Mercury-Atlas 7)	1962 Tau 1	24 May 1962	Atlas D
Tiros 5	1962 Alpha Alpha	19 June 1962	Thor-Delta
Telstar 1	1962 Alpha Epsilon 1	10 July 1962	Thor-Delta

Name	International Designation*	Launch Date**	Launch Vehicle***
Echo (AVT–2)	(suborbital)	18 July 1962	Thor
Mariner 2	1962 Alpha Rho 1	27 August 1962	Atlas-Agena B
Tiros 6	1962 Alpha Psi 1	18 September 1962	Thor-Delta
Alouette 1	1962 Beta Alpha 1	28 September 1962	Thor-Agena B
Explorer 14	1962 Beta Gamma 1	2 October 1962	Thor-Delta
Sigma 7 (Mercury-Atlas 8)	1962 Beta Delta 1	3 October 1962	Atlas D
Ranger 5	1962 Beta Eta 1	18 October 1962	Atlas-Agena B
Explorer 15	1962 Beta Lambda 1	27 October 1962	Thor-Delta
Saturn-Apollo 3 (SA–3)	(suborbital)	16 November 1962	Saturn I
Relay 1	1962 Beta Upsilon 1	13 December 1962	Thor-Delta
Explorer 16	1962 Beta Chi 1	16 December 1962	Scout
Syncom 1	1963–4A	14 February 1963	Thor-Delta
Saturn-Apollo 4 (SA–4)	(suborbital)	28 March 1963	Saturn I
Explorer 17	1963–9A	2 April 1963	Thor-Delta
Telstar 2	1963–13A	7 May 1963	Thor-Delta
Faith 7 (Mercury-Atlas 9)	1963–15A	15 May 1963	Atlas D
Tiros 7	1963–24A	19 June 1963	Thor-Delta
Syncom 2	1963–31A	26 July 1963	Thor-Delta
Little Joe II	(suborbital)	28 August 1963	Little Joe II
Explorer 18	1963–46A	26 November 1963	Thor-Delta
Atlas-Centaur 2	1963–47A	27 November 1963	Atlas-Centaur
Explorer 19	1963–53A	19 December 1963	Scout
Tiros 8	1963–54A	21 December 1963	Thor-Delta
Relay 2	1964–3A	21 January 1964	Thor-Delta
Echo 2	1964–4A	25 January 1964	Thor-Agena B
Saturn-Apollo 5 (SA–5)	1964–5A	29 January 1964	Saturn I

MAJOR NASA LAUNCH RECORD

Name	International Designation*	Launch Date**	Launch Vehicle***
Ranger 6	1964–7A	30 January 1964	Atlas-Agena B
Ariel 2	1964–15A	27 March 1964	Scout
Gemini-Titan 1	1964–18A	8 April 1964	Gemini-Titan II
FIRE 1	(suborbital)	14 April 1964	Atlas D
Apollo (A–001) Transonic Abort	(suborbital)	13 May 1964	Little Joe II
Apollo-Saturn 101 (SA–6)	1964–25A	28 May 1964	Saturn I
Atlas-Centaur 3	(suborbital)	30 June 1964	Atlas-Centaur
SERT 1	(suborbital)	20 July 1964	Scout
Ranger 7	1964–41A	28 July 1964	Atlas-Agena B
Reentry 4	(suborbital)	18 August 1964	Scout
Syncom 3	1964–47A	19 August 1964	TAT–Delta
Explorer 20	1964–51A	25 August 1964	Scout
Nimbus 1	1964–52A	28 August 1964	Thor-Agena B
OGO 1	1964–54A	4 September 1964	Atlas-Agena B
Apollo-Saturn 102 (SA–7)	1964–57A	18 September 1964	Saturn I
Explorer 21	1964–60A	4 October 1964	Thor-Delta
Explorer 22	1964–64A	9 October 1964	Scout
Mariner 3	1964–73A	5 November 1964	Atlas-Agena D
Explorer 23	1964–74A	6 November 1964	Scout
Explorer 24 and	1964–76A	21 November 1964	Scout
Explorer 25 *(Injun 4)*	1964–76B		
Mariner 4	1964–77A	28 November 1964	Atlas-Agena D
Apollo Maximum Q Abort	(suborbital)	8 December 1964	Little Joe II
Atlas-Centaur 4	1964–82A	11 December 1964	Atlas-Centaur
San Marco 1	1964–84A	15 December 1964	Scout
Explorer 26	1964–86A	21 December 1964	Thor-Delta

Name	International Designation*	Launch Date**	Launch Vehicle***
Gemini-Titan 2	(suborbital)	19 January 1965	Titan II
Tiros 9	1965–4A	22 January 1965	Thor-Delta
OSO 2	1965–7A	3 February 1965	Thor-Delta
Pegasus 1 and	1965–9A	16 February 1965	Saturn I
Apollo-Saturn 103 (SA–9)	1965–9B		
Ranger 8	1965–10A	17 February 1965	Atlas-Agena B
Ranger 9	1965–23A	21 March 1965	Atlas-Agena B
Gemini 3 (Gemini-Titan 3)	1965–24A	23 March 1965	Titan II
Early Bird 1	1965–28A	6 April 1965	TAT–Delta
Explorer 27	1965–32A	29 April 1965	Scout
FIRE 2	(suborbital)	22 May 1965	Atlas D
Pegasus 2 and	1965–39A	25 May 1965	Saturn I
Apollo-Saturn 104 (SA–8)	1965–39B		
Explorer 28	1965–42A	29 May 1965	Thor-Delta
Gemini 4 (Gemini-Titan 4)	1965–43A	3 June 1965	Titan II
Tiros 10	1965–51A	2 July 1965	TAT–Delta
Pegasus 3 and	1965–60A	30 July 1965	Saturn I
Apollo-Saturn 105 (SA–10)	1965–60B		
Scout (SEV–A)	1965–63A	10 August 1965	Scout
Centaur-Surveyor (Atlas-Centaur 6)	1965–64A	11 August 1965	Atlas-Centaur
Gemini 5 (Gemini-Titan 5)	1965–68A	21 August 1965	Titan II
OGO 2	1965–81A	14 October 1965	TAT–Agena D
Explorer 29 (*GEOS 1*)	1965–89A	6 November 1965	TAID

NASA MAJOR LAUNCH RECORD

Name	International Designation*	Launch Date**	Launch Vehicle***
Explorer 30	1965–93A	18 November 1965	Scout
Alouette 2 and	1965–98A	28 November 1965	Thor-Agena B
Explorer 31	1965–98B		
Gemini 7 (Gemini-Titan 7)	1965–100A	4 December 1965	Titan II
FR–1	1965–101A	6 December 1965	Scout
Gemini 6 (Gemini-Titan 6)	1965–104A	15 December 1965	Titan II
Pioneer 6	1965–105A	16 December 1965	TAID
Apollo (A–004) Intermediate Altitude Abort	(suborbital)	20 January 1966	Little Joe II
ESSA 1	1966–8A	3 February 1966	Thor-Delta
Reentry 5	(suborbital)	9 February 1966	Scout
Apollo-Saturn 201	(suborbital)	26 February 1966	Saturn IB
ESSA 2	1966–16A	28 February 1966	TAID
Gemini-Agena Target Vehicle 8	1966–19A	16 March 1966	Atlas-Agena D
Gemini 8 (Gemini-Titan 8)	1966–20A	16 March 1966	Titan II
Centaur-Surveyor (Atlas-Centaur 8)	1966–30A	7 April 1966	Atlas-Centaur
OAO 1	1966–31A	8 April 1966	Atlas-Agena D
Nimbus 2	1966–40A	15 May 1966	TAT–Agena B
Explorer 32	1966–44A	25 May 1966	Thor-Delta
Surveyor 1	1966–45A	30 May 1966	Atlas-Centaur
Augmented Target Docking Adapter	1966–46A	1 June 1966	Atlas D
Gemini 9 (Gemini-Titan 9)	1966–47A	3 June 1966	Titan II
OGO 3	1966–49A	6 June 1966	Atlas-Agena B
PAGEOS 1	1966–56A	23 June 1966	TAT–Agena D
Explorer 33	1966–58A	1 July 1966	TAID

Name	International Designation*	Launch Date**	Launch Vehicle***
Apollo-Saturn 203	1966–59A	5 July 1966	Saturn IB
Gemini-Agena Target Vehicle 10	1966–65A	18 July 1966	Atlas-Agena D
Gemini 10 (Gemini-Titan 10)	1966–66A	18 July 1966	Titan II
Lunar Orbiter 1	1966–73A	10 August 1966	Atlas-Agena D
Pioneer 7	1966–75A	17 August 1966	TAID
Apollo-Saturn 202	(suborbital)	25 August 1966	Saturn IB
Gemini-Agena Target Vehicle 11	1966–80A	12 September 1966	Atlas-Agena D
Gemini 11 (Gemini-Titan 11)	1966–81A	12 September 1966	Titan II
Surveyor 2	1966–84A	20 September 1966	Atlas-Centaur
ESSA 3	1966–87A	2 October 1966	TAID
Centaur-Surveyor (Atlas-Centaur 9)	1966–95A	25 October 1966	Atlas-Centaur
Intelsat-II F–1 (Intelsat II–A)	1966–96A	27 October 1966	TAID
Lunar Orbiter 2	1966–100A	6 November 1966	Atlas-Agena D
Gemini-Agena Target Vehicle 12	1966–103A	11 November 1966	Atlas-Agena D
Gemini 12 (Gemini-Titan 12)	1966–104A	11 November 1966	Titan II
ATS 1	1966–110A	6 December 1966	Atlas-Agena D
Biosatellite 1	1966–114A	14 December 1966	TAID
Intelsat-II F–2	1967–1A	11 January 1967	TAID
ESSA 4	1967–6A	26 January 1967	TAID
Lunar Orbiter 3	1967–8A	4 February 1967	Atlas-Agena D
OSO 3	1967–20A	8 March 1967	Thor-Delta
Intelsat-II F–3	1967–26A	22 March 1967	TAID
ATS 2	1967–31A	5 April 1967	Atlas-Agena D
Surveyor 3	1967–35A	17 April 1967	Atlas-Centaur

NASA MAJOR LAUNCH RECORD

Name	International Designation*	Launch Date**	Launch Vehicle***
ESSA 5	1967–36A	20 April 1967	TAID
San Marco 2	1967–38A	26 April 1967	Scout
Lunar Orbiter 4	1967–41A	4 May 1967	Atlas-Agena D
Ariel 3	1967–42A	5 May 1967	Scout
Explorer 34	1967–51A	24 May 1967	TAID
Mariner 5	1967–60A	14 June 1967	Atlas-Agena D
Surveyor 4	1967–68A	14 July 1967	Atlas-Centaur
Explorer 35	1967–70A	19 July 1967	TAID
OGO 4	1967–73A	28 July 1967	TAT–Agena D
Lunar Orbiter 5	1967–75A	1 August 1967	Atlas-Agena D
Biosatellite 2	1967–83A	7 September 1967	TAID
Surveyor 5	1967–84A	8 September 1967	Atlas-Centaur
Intelsat-II F–4	1967–94A	27 September 1967	TAID
OSO 4	1967–100A	18 October 1967	Thor-Delta
RAM C–1	(suborbital)	19 October 1967	Scout
ATS 3	1967–111A	5 November 1967	Atlas-Agena D
Surveyor 6	1967–112A	7 November 1967	Atlas-Centaur
Apollo 4 (Apollo-Saturn 501)	1967–113A	9 November 1967	Saturn V
ESSA 6	1967–114A	10 November 1967	TAID
Pioneer 8 and	1967–123A	13 December 1967	TAID
TTS 1	1967–123B		
Surveyor 7	1968–1A	7 January 1968	Atlas-Centaur
Explorer 36 (GEOS 2)	1968–2A	11 January 1968	TAID
Apollo 5 (Apollo-Saturn 204)	1968–7A	22 January 1968	Saturn IB
OGO 5	1968–14A	4 March 1968	Atlas-Agena D
Explorer 37	1968–17A	5 March 1968	Scout
Apollo 6 (Apollo-Saturn 502)	1968–25A	4 April 1968	Saturn V

Name	International Designation*	Launch Date**	Launch Vehicle***
Reentry 6	(suborbital)	27 April 1968	Scout
IRIS 1	1968–41A	16 May 1968	Scout
Explorer 38	1968–55A	4 July 1968	TAID
Explorer 39 and	1968–66A	8 August 1968	Scout
Explorer 40 (Injun 5)	1968–66B		
ATS 4	1968–68A	10 August 1968	Atlas-Centaur
ESSA 7	1968–69A	16 August 1968	LTTAT–Delta
RAM C–2	(suborbital)	22 August 1968	Scout
Aurorae	1968–84A	3 October 1968	Scout
Apollo 7 (Apollo-Saturn 205)	1968–89A	11 October 1968	Saturn IB
Pioneer 9 and	1968–100A	8 November 1968	TAID
TETR 2	1968–100B		
HEOS 1	1968–109A	5 December 1968	TAID
OAO 2	1968–110A	7 December 1968	Atlas-Centaur
ESSA 8	1968–114A	15 December 1968	LTTAT–Delta
Intelsat-III F–2	1968–116A	18 December 1968	LTTAT–Delta
Apollo 8 (Apollo-Saturn 503)	1968–118A	21 December 1968	Saturn V
OSO 5	1969–6A	22 January 1969	Thor-Delta
ISIS 1	1969–9A	30 January 1969	TAID
Intelsat-III F–3	1969–11A	5 February 1969	LTTAT–Delta
Mariner 6	1969–14A	24 February 1969	Atlas-Centaur
ESSA 9	1969–16A	26 February 1969	TAID
Apollo 9 (Apollo-Saturn 504)	1969–18A	3 March 1969	Saturn V
Mariner 7	1969–30A	27 March 1969	Atlas-Centaur
Nimbus 3	1969–37A	14 April 1969	LTTAT–Agena D

NASA MAJOR LAUNCH RECORD

Name	International Designation*	Launch Date**	Launch Vehicle***
Apollo 10 (Apollo-Saturn 505)	1969–43A	18 May 1969	Saturn V
Intelsat-III F–4	1969–45A	21 May 1969	LTTAT–Delta
OGO 6	1969–51A	5 June 1969	LTTAT–Agena D
Explorer 41	1969–53A	21 June 1969	TAID
Biosatellite 3	1969–56A	28 June 1969	LTTAT–Delta
Apollo 11 (Apollo-Saturn 506)	1969–59A	16 July 1969	Saturn V
Intelsat-III F–5	1969–64A	25 July 1969	LTTAT–Delta
OSO 6 and *PAC*	1969–68A 1969–68B	9 August 1969	LTTAT–Delta
ATS 5	1969–69A	12 August 1969	Atlas-Centaur
Boreas	1969–83A	1 October 1969	Scout
Azur	1969–97A	7 November 1969	Scout
Apollo 12 (Apollo-Saturn 507	1969–99A	14 November 1969	Saturn V
Skynet A	1969–101A	21 November 1969	LTTAT–Delta
Intelsat-III F–6	1970–3A	14 January 1970	LTTAT–Delta
ITOS 1 and *OSCAR 5*	1970–8A 1970–8B	23 January 1970	LTTAT–Delta
SERT 2	1970–9A	3 February 1970	LTTAT–Agena D
NATOSAT 1	1970–21A	20 March 1970	LTTAT–Delta
Nimbus 4	1970–25A	8 April 1970	LTTAT–Agena D
Apollo 13 (Apollo-Saturn 508)	1970–29A	11 April 1970	Saturn V
Intelsat-III F–7	1970–32A	22 April 1970	LTTAT–Delta
Intelsat-III F–8	1970–55A	23 July 1970	LTTAT–Delta
Skynet B	1970–62A	19 August 1970	LTTAT–Delta
RAM C–3	(suborbital)	30 September 1970	Scout

Name	International Designation*	Launch Date**	Launch Vehicle***
OFO	1970–94A	9 November 1970	Scout
and			
RM	1970–94B		
NOAA 1	1970–106A	11 December 1970	LTTAT–Delta
and			
CEPE	1970–106B		
Explorer 42	1970–107A	12 December 1970	Scout
Intelsat-IV F–2	1971–6A	25 January 1971	Atlas-Centaur
Apollo 14 (Apollo-Saturn 509)	1971–8A	31 January 1971	Saturn V
NATOSAT 2	1971–9A	2 February 1971	LTTAT–Delta
Explorer 43	1971–19A	13 March 1971	TAID
ISIS 2	1971–24A	31 March 1971	TAID
San Marco 3	1971–36A	24 April 1971	Scout
Mariner 9	1971–51A	30 May 1971	Atlas-Centaur
PAET	(suborbital)	20 June 1971	Scout
Explorer 44	1971–58A	8 July 1971	Scout
Apollo 15 (Apollo-Saturn 510)	1971–63A	26 July 1971	Saturn V
Eole	1971–71A	16 August 1971	Scout
OSO 7	1971–83A	29 September 1971	LTTAT–Delta
and			
TETR 3	1971–83B		
Explorer 45	1971–96A	15 November 1971	Scout
Ariel 4	1971–109A	11 December 1971	Scout
Intelsat-IV F–3	1971–116A	19 December 1971	Atlas-Centaur
Intelsat-IV F–4	1972–3A	22 January 1972	Atlas-Centaur
HEOS 2	1972–5A	31 January 1972	LTTAT–Delta
Pioneer 10	1972–12A	2 March 1972	Atlas-Centaur-TE–M–364–4
TD–1A	1972–14A	11 March 1972	LTTAT–Delta
Apollo 16 (Apollo-Saturn 511)	1972–31A	16 April 1972	Saturn V

NASA MAJOR LAUNCH RECORD

Name	International Designation*	Launch Date**	Launch Vehicle***
Intelsat-IV F–5	1972–41A	13 June 1972	Atlas-Centaur
ERTS 1	1972–58A	23 July 1972	LTTAT–Delta
Explorer 46	1972–61A	13 August 1972	Scout
OAO 3	1972–65A	21 August 1972	Atlas-Centaur
Triad OI–1X	1972–69A	2 September 1972	Scout
Explorer 47	1972–73A	22 September 1972	LTTAT–Delta
NOAA 2	1972–82A	15 October 1972	LTTAT–Delta
and			
OSCAR 6	1972–82B		
Anik 1	1972–90A	9 November 1972	LTTAT–Delta
Explorer 48	1972–91A	16 November 1972	Scout
ESRO 4	1972–92A	21 November 1972	Scout
Apollo 17 (AS–512)	1972–96A	7 December 1972	Saturn V
Nimbus 5	1972–97A	10 December 1972	LTTAT–Delta
Aeros	1972–100A	16 December 1972	Scout
Pioneer 11	1973–19A	5 April 1973	Atlas-Centaur-TE–M–364–4
Anik 2	1973–23A	20 April 1973	LTTAT–Delta
Skylab 1	1973–27A	14 May 1973	Saturn V
Skylab 2	1973–32A	25 May 1973	Saturn IB
Explorer 49	1973–39A	10 June 1973	LTTAT–Delta
Skylab 3	1973–50A	28 July 1973	Saturn IB
Intelsat-IV F–7	1973–58A	23 August 1973	Atlas-Centaur
Explorer 50	1973–78A	25 October 1973	LTTAT–Delta
NNSS O–20	1973–81A	29 October 1973	Scout
Mariner 10	1973–85A	3 November 1973	Atlas-Centaur
NOAA 3	1973–86A	6 November 1973	LTTAT–Delta
Skylab 4	1973–90A	16 November 1973	Saturn IB
Explorer 51	1973–101A	15 December 1973	LTTAT–Delta
Skynet IIA	1974–2A	18 January 1974	LTTAT-Delta

Name	International Designation*	Launch Date**	Launch Vehicle***
San Marco 4	1974–9A	18 February 1974	Scout
Miranda (UK–X4)	1974–13A	8 March 1974	Scout
Westar 1	1974–22A	13 April 1974	LTTAT–Delta
SMS 1	1974–33A	17 May 1974	LTTAT–Delta
ATS 6	1974–39A	30 May 1974	Titan IIIC
Hawkeye 1 (Explorer 52)	1974–40A	3 June 1974	Scout
Aeros 2	1974–55A	16 July 1974	Scout
ANS	1974–70A	30 August 1974	Scout
Westar 2	1974–75A	10 October 1974	LTTAT–Delta
Ariel 5 (UK–5)	1974–77A	15 October 1974	Scout
NOAA 4	1974–89A	15 November 1974	LTTAT–Delta
and *OSCAR 7*	1974–89B		
and *INTASAT*	1974–89C		
Intelsat-IV F–8	1974–93A	21 November 1974	Atlas-Centaur
Skynet IIB	1974–94A	23 November 1974	LTTAT–Delta
Helios 1	1974–97A	10 December 1974	Titan IIIE–Centaur-TE–M–364–4
Symphonie 1	1974–101A	18 December 1974	LTTAT–Delta

APPENDIX D

NASA NAMING COMMITTEES

The first "naming committee" established within NASA Headquarters was the Ad Hoc Committee to Name Space Projects and Objects. Meeting informally during 1960, the Committee sought to establish procedures for submitting and selecting names and proposed specific mission categories as a step toward defining a clear-cut pattern of NASA names. Precedent had been set for the continuation of a "series approach" to names by Explorer and Pioneer spacecraft. The Committee emphasized that flight names should be suggestive of the mission and reflect the series of which they were a part. This emphasis was the basis for the decision to use the "Cortright" system for naming space probes, as described in the introductory section on space probes in the text (Part III).

On 9 January 1961, NASA Management Instruction 4-3-1, by the Committee, prescribed policy and procedure for assigning names to major NASA projects. A new committee, the Project Designation Committee, would be appointed to review and recommend specific project names. The NMI stated in part:

> Each project name will be a simple euphonic word that will not duplicate or be confused with other NASA or non-NASA project titles. When possible and if appropriate, names will be chosen to reflect NASA's mission. Project names will be serialized when appropriate, thus limiting the number of different names in use at any one time; however, serialization will be used only after successful flight or accomplishment has been achieved.

The Project Designation Committee met to consider specific names and solicited suggestions from NASA field centers as categories for future mission names were defined. Names chosen were "reserved" for the appropriate missions. The committee recommendations were not always approved and the selection of a name for a particular mission occasionally was postponed for lack of an acceptable substitute. Many approved names were never used, as the projects themselves were redesigned or later canceled.

The influence of the committee waned after 1963 as some projects were deferred or canceled and ongoing series required no new names. Fewer new projects were approved and recommendation and approval of project names often came after the fact; names already in common use by program offices were adopted. Revived in 1970, the committee meets only to consider specific requests for official project names.

REFERENCE NOTES

PART I: LAUNCH VEHICLES

ABLE

1. Milton W. Rosen, Office of Defense Affairs, NASA, telephone interview, 16 Feb. 1965.

AGENA

1. W. F. Whitmore, "AGENA—The Spacecraft and the Star," Lockheed Missiles & Space Co. research paper, 16 Jan. 1969; and Dick Bissinette, Andrews AFB, letter to Judy Gildea, NASA, 27 March 1963.
2. *Ibid.*; and R. H. Allen, *Star Names: Their Lore and Meaning* (New York: Dover Publications, Inc., 1963), p. 154. The Dover edition is an unabridged and corrected republication of the work first published by G. E. Steckert in 1899 under the title *Star-Names and Their Meanings.*
3. D. H. Menzel, footnote in Allen, *Star Names*, p. 154; and Whitmore, "AGENA—The Spacecraft and the Star."
4. R. Cargill Hall, Lockheed Missiles & Space Co., letter to Historical Staff, NASA, 26 Aug. 1965.
5. Robert F. Piper, Historical Office, Air Force Space Systems Division, letter to Historical Staff, NASA, 31 Aug. 1965.
6. Hall, "The Agena Satellite" (unpublished essay), November 1966.

ATLAS

1. John L. Chapman, *Atlas: The Story of a Missile* (New York: Harper & Bros., 1960), p. 62; and R. T. Blair, Jr., Convair Division, General Dynamics Corp., letter to Historical Staff, NASA, 10 Sept. 1965.
2. Robert F. Piper, Historical Office, Air Force Space Systems Division, letter to Historical Staff, NASA, 31 Aug. 1965; B/G D. N. Yates, Director of Research and Development, Office of the Deputy Chief of Staff for Development, Hq. USAF, memorandum to the Chairman, Committee on Guided Missiles, Research and Development Board, DOD, 30 July 1951; S. D. Cornell, Acting Executive Secretary, Committee on Guided Missiles, memorandum to B/G D. N. Yates, 6 Aug. 1951; and Yates, memorandum to Commanding General, Air Research and Development Command, 27 Aug. 1951.
3. NASA. News Release 75-19.

BIG JOE

1. Loyd S. Swenson, Jr., James M. Grimwood, and Charles C. Alexander, *This New Ocean: A History of Project Mercury,* NASA SP–4201 (Washington: NASA, 1966), p. 125; and Paul E. Purser, Manned Spacecraft Center (MSC), handwritten note to James M. Grimwood, Historian, MSC [October 1963].

CENTAUR

1. Frank Kerr, Astronautics Division, General Dynamics Corp., teletype message to Lynn Manley, Lewis Research Center (LeRC), 10 Dec. 1963; and Lynn Manley, letter to Historical Staff, NASA, 11 Dec. 1963.

DELTA

1. L. B. Norris, Goddard Space Flight Center (GSFC), letter to Historical Staff, NASA, 10 Dec. 1963; and Robert L. Perry, "The Atlas, Thor, Titan, and Minuteman," in Eugene M. Emme, ed., *The History of Rocket Technology* (Detroit: Wayne State University Press, 1964), p. 160.

2. Milton W. Rosen, Office of Defense Affairs, NASA, telephone interview 16 Feb. 1965.

3. NASA, News Releases 64–133 and 67–306; NASA, program office; and Communications Satellite Corp., Press Kit: INTELSAT III [1968].

4. NASA, News Releases 70–2, 72–206, and 74–77K; Robert J. Goss, "Delta Vehicle Improvements" in *Significant Accomplishments in Technology,* proceedings of symposium at Goddard Space Flight Center, 7–8 Nov. 1972, NASA SP–326 (Washington: NASA, 1973), 11–13.

5. NASA, News Releases 60–237, 60–242, 74–77K, and 75–19.

JUNO

1. William H. Pickering, Director, Jet Propulsion Laboratory (JPL), teletype message to M/G John B. Medaris, Commanding General, ABMA, 18 Nov. 1957; Medaris, teletype message to Gen. James M. Gavin, Chief of Research and Development, Hq. USA, 20 Nov. 1957; and Wernher von Braun, "The Redstone, Jupiter, and Juno," in Eugene M. Emme, ed., *The History of Rocket Technology* (Detroit: Wayne State University Press, 1964), pp. 107–121.

2. R. Cargill Hall, JPL, letter to Eugene M. Emme, Historian, NASA, 9 Oct. 1965.

3. Von Braun, "The Redstone, Jupiter, and Juno."

LITTLE JOE

1. Loyd S. Swenson, Jr., James M. Grimwood, and Charles C. Alexander, *This New Ocean: A History of Project Mercury,* NASA SP–4201 (Washington: NASA, 1966), pp. 123–124; Paul E. Purser, MSC, handwritten note to James M. Grimwood, Historian, MSC [October 1963]; and Robert W. Mulac, Langley Research Center (LaRC), letter to Historical Staff, NASA, 10 Dec. 1963.

REDSTONE

1. Wernher von Braun, "The Redstone, Jupiter, and Juno," in Eugene M. Emme, ed., *The History of Rocket Technology* (Detroit: Wayne State University Press, 1964), p. 109.

2. David S. Akens, "Historical Sketch of Marshall Space Flight Center" (MS), n.d. Redstone Arsenal was so named 26 Feb. 1943, having been designated "Redstone Ordnance Plant" since 6 Oct. 1941. From 4 Aug. 1941 to 6 Oct. 1941 it was called "Huntsville Arsenal."

SATURN I, SATURN IB

1. Wernher von Braun, "The Redstone, Jupiter, and Juno," in Eugene M. Emme, ed., *The History of Rocket Technology* (Detroit: Wayne State University Press, 1964), p. 119.

2. A. Ruth Jarrell, Historical Office, Marshall Space Flight Center (MSFC), letter to Historical Staff, NASA, 16 Dec. 1963; B/G J. A. Barclay, Commander, ABMA, memorandum to M/G John B. Medaris, Commanding General, Army Ordnance Missile Command, 15 Oct. 1958; Medaris, teletype message to Gen. James M. Gavin, Chief of Research and Development, Hq. USA, 15 Oct. 1958; and MSFC, Historical Office, *Saturn Illustrated Chronology*, MHR-4 (Huntsville: MSFC, 1965), p. 5.

3. Silverstein *et al.*, "Report to the Administrator, NASA, on Saturn Development Plan by Saturn Vehicle Team," 15 Dec. 1959; and MSFC, Historical Office, *Saturn Illustrated Chronology*, pp. 8–9. The C–1 configuration was composed of three stages: S–1—the first stage, with 6.7-million-newton (1.5-million-pound) thrust—and S–IV and S–V liquid-hydrogen-powered second and third stages. The S–V stage was eliminated in May 1961.

4. MSFC, Historical Office, *Saturn Illustrated Chronology*, p. 56; and MSFC, Historical Office, *History of the George C. Marshall Space Flight Center: January 1–June 30, 1962*, MHM–5, 1 (Huntsville: MSFC, 1962), 28. The decision to develop the C–1B was announced 11 July 1962, but the name had been in use since early 1962 in design and feasibility studies.

5. George L. Simpson, Jr., Assistant Administrator for Public Affairs, NASA, memorandum for the Associate Administrator, NASA, 7 Jan. 1963; and MSFC, Historical Office, *Saturn Illustrated Chronology*, p. 69.

6. Julian W. Scheer, Assistant Administrator for Public Affairs, NASA, memorandum from Project Designation Committee to Public Information Director *et al.*, NASA, 9 June 1966.

7. Robert C. Seamans, Jr., Deputy Administrator, NASA, memorandum for George E. Mueller, Associate Administrator for Manned Space Flight, NASA, 6 Jan. 1967; Willis H. Shapley, Associate Deputy Administrator, NASA, memorandum for Mueller *et al.*, NASA, 2 Dec. 1967; and Scheer, memorandum to O. B. Lloyd, Jr., Public Information Director, *et al.*, NASA, 15 Jan. 1968.

SATURN V

1. MSFC, Historical Office, *History of the George C. Marshall Space Flight Center: July 1–December 31, 1961*, MHM–4, 1 (Huntsville: MSFC, 1962), 33.

2. George L. Simpson, Jr., Assistant Administrator for Public Affairs, NASA, memorandum for the Associate Administrator, NASA, 7 Jan. 1963. This memo recommended name changes for all three Saturn vehicles—for Saturn I, Saturn IB, and Saturn V.

3. MSFC, Historical Office, *Saturn Illustrated Chronology*, MHR–4 (Huntsville: MSFC, 1965), p. 69.

4. NASA, News Release 72–220K.

SCOUT

1. William E. Stoney, Jr., MSC, letter to Historical Staff, NASA, 13 April 1965.

2. NASA, News Release 75–19.

3. NASA, Wallops Station, Open House Program, 29–30 Sept. 1963.

4. NASA, News Release 74–138; and Paul E. Goozh, Scout Program Manager, NASA, telephone interview, 17 June 1974.

SHOTPUT

1. NASA, News Releases 60–158 and 60–186. The second stage of the Shotput vehicle was an Allegany Ballistics Laboratory X–248 rocket, originally designed for the Vanguard and Thor-Able vehicles.

2. William J. O'Sullivan, Jr., Head of the Space Vehicle Group, LaRC, letter to Don Murray, 29 Sept. 1960.

THOR

1. W. C. Cleveland, Director of Public Relations, Douglas Missile & Space Systems Division, letter to Historical Staff, NASA, 27 Aug. 1965.

2. Thomas Bulfinch, *Mythology,* Edmund Fuller, ed. (New York: Dell Publishing Co., Inc., 1959), p. 243.

3. Lloyd Mallan, *Peace Is a Three-Edged Sword* (New York: Prentice-Hall [1964]), pp. 190–192.

4. NASA, News Release 68–84.

5. *Ibid.*; USAF, News Release 205.65; and NASA, program office.

TITAN

1. R. L. Tonsing, Director of Public Relations, Aerospace Division, Martin Marietta Corp., letter to Historical Staff, NASA, 26 Sept. 1965; and Lloyd Mallan, *Peace Is a Three-Edged Sword* (New York: Prentice-Hall [1964]), pp. 190–192.

2. Thomas Bulfinch, *Mythology,* Edmund Fuller, ed. (New York: Dell Publishing Co., Inc., 1959), p. 15.

3. NASA, News Release 75–19.

PART II: SATELLITES

1. See *Webster's Third New International Dictionary* (unabridged); and Woodford Agee Heflin, *The Second Aerospace Glossary* (Maxwell AFB, Ala.: Air University, 1966).

AEROS

1. NASA, News Release 69–91. The memorandum of understanding was signed 10 June 1969.

2. Lloyd E. Jones, Jr., Office of International Affairs, NASA, telephone interview, 4 June 1971.

3. NASA, "Project Approval Document," 27 Feb. 1970; NASA, program office.

ALOUETTE

1. See Jonathan D. Caspar, "The Alouette (S–27) Program: A Case Study in NASA International Cooperative Activities," HHN–42, 1964 (comment ed.) and 1965 (revised MS).

2. N. W. Morton, Dept. of National Defence, Canadian Joint Staff, letter to Arnold W. Frutkin, Director of International Programs, NASA, 27 April 1961.

3. Robert C. Seamans, Jr., Associate Administrator, NASA, letter to N. W. Morton, 11 May 1961; and NASA, Announcement 312, 24 May 1961.

4. NASA, News Release 64–207; and Wallops Station, News Release 64–77.

ANS

1. NASA, News Release 70–91.

ARIEL

1. D. J. Gerhard, Office of the Scientific Attaché, U.K. Scientific Mission, Washington, D.C., letter to Arnold W. Frutkin, Director of International Programs, NASA, 15 Dec. 1961; Frutkin, memorandum to Robert C. Seamans, Jr., Associate Administrator, NASA, 15 Dec. 1961; and Boyd C. Myers II (Chairman, NASA Project Designation Committee), Director, Program Review and Resources Management, NASA, memorandum to Seamans, 2 Feb. 1962, with approval signature of Dr. Seamans.

2. NASA, News Release 74–36; NASA program office; UN Document A/AC.105/INF.289, 18 April 1974.

ATS

1. U.S. Congress, Senate, Committee on Aeronautical and Space Sciences, *Hearings . . . NASA Authorization for FY 1964,* Pt. 1, April 1963 (Washington: 1963), pp. 8, 143, 433–434.

2. Homer E. Newell, Associate Administrator for Space Science and Applications, NASA, in U.S. Congress, Senate, Committee on Aeronautical and Space Sciences, *Hearings . . . NASA Authorization for FY 1965,* Pt. 2, March 1964 (Washington: 1964), p. 559.

3. NASA, News Release 64–50.

4. Robert F. Garbarini, Director of Engineering, Office of Space Science and Applications, NASA, memorandum to Director, Communication and Navigation Programs Division, NASA, 11 Sept. 1964.

5. Newell, memorandum to Robert C. Seamans, Jr., Associate Administrator, NASA, 2 Oct. 1964, with approval signature of Dr. Seamans.

AZUR

1. "Memorandum of Understanding between the German Ministry for Scientific Research and the United States National Aeronautics and Space Administration," attachment to NASA, News Release 69–146.

2. Charles F. Rice, Jr., GSFC [former AZUR Project Coordinator at GSFC], telephone interview, 2 June 1971.

BIOSATELLITE

1. NASA, News Release 66–312.

2. O. E. Reynolds, Director of Bioscience Programs, NASA, memorandum to Harold L. Goodwin (Member, NASA Project Designation Committee), Director, Office of Program

Development, NASA, 4 June 1962; and Jack Posner, Office of Space Science and Applications, NASA, telephone interview, 10 Aug. 1965.

3. NASA Project Designation Committee, minutes of meeting 9 Jan. 1963.

ECHO

1. William J. O'Sullivan, "Notes on Project Echo" (MS), n.d.; and NASA, News Release 61-252.

2. John R. Pierce, "Orbital Radio Relays," *Jet Propulsion,* 25 (April 1955), 153–157.

3. E. W. Morse, "Preliminary History of the Origins of Project Syncom" (MS), 1964.

4. Robert W. Mulac, LaRC, letter to Historical Staff, NASA, 10 Dec. 1963.

EOLE

1. NASA, News Release 66–156.

2. *Ibid.*

3. NASA, "Project Approval Document," 7 Dec. 1966; NASA, "Research and Development: Cooperative Effort/Flight" [FY 1969 Project Approval Document], 2 Dec. 1968; and NASA, News Release 70–222.

ERTS

1. John Hanessian, Jr., "International Aspects of Earth Resources Survey Satellite Programs," *Journal of the British Interplanetary Society,* 23 (Spring 1970), 535, 541.

2. Although development funds were not approved until FY 1970, the first Project Approval Document was dated 7 Jan. 1969.

3. Eldon D. Taylor, Director of Program Review and Resources Management, NASA, memorandum to Chief, Management Issuances Section, NASA, 7 July 1966; and W. Fred Boone, Assistant Administrator for Defense Affairs, NASA, memorandum to Leonard Jaffe, Director of Space Applications Programs, NASA, 22 July 1966.

4. John E. Naugle, Associate Administrator for Space Science and Applications, NASA, prepared statement in U.S. Congress, House of Representatives, Committee on Science and Astronautics' Subcommittee on Space Science and Applications, *Hearings. . . 1972 NASA Authorization,* Pt. 3, March 1971 (Washington: 1971), pp. 156–157.

5. Jaffe, memorandum to Julian W. Scheer, Assistant Administrator for Public Affairs, NASA, 10 Jan. 1967.

6. Department of the Interior, Office of the Secretary, release, "Earth's Resources to Be Studied from Space," 21 Sept. 1966.

7. NASA, Program Review and Resources Management Office, "Chronology of NASA Earth Resources Program," 13 Dec. 1971; and George J. Vecchietti, Director of Procurement, NASA, memorandum to Philip N. Whittaker, Assistant Administrator for Industry Affairs, NASA, 14 Oct. 1968.

8. James A. Long, Assistant Executive Secretary, NASA, notes of Project Designation Committee meeting, 17 Feb. 1970; and Long, telephone interview, 15 April 1971.

9. U.S. Congress, House of Representatives, Committee on Science and Astronautics' Subcommittee on Space Science and Applications, *Hearings. . . 1975 NASA Authorization,* Pt. 3, February and March 1974 (Washington: 1974), pp. 50, 91; NASA, Project Approval Document, Encl. 27 to 61–600-SA, 14 Jan. 1974; and NASA, News Release 74–80.

10. *Hearings. . . 1975 NASA Authorization,* Pt. 3, p. 50.

ESRO

1. NASA, News Release 64–178; and Ellen T. Rye, Office of International Affairs, NASA, telephone interview, 20 April 1967.

2. "Memorandum of Understanding between the European Space Research Organization and the United States National Aeronautics and Space Administration," attachment to NASA, News Release 64–178; and NASA, News Release, "Press Briefing: ESRO II and NASA International Cooperative Programs," 19 May 1967.

3. NASA, program office; and NASA, News Release 68–158.

4. NASA, program office.

5. "Memorandum of Understanding."

6. NASA, News Releases 66–332 and 68–204.

7. European Space Research Organization, *Europe in Space,* March 1974, pp. 20–41; and A. V. Cleaver, *Spaceflight,* 16, No. 6 (June 1974), 220–237.

ESSA

1. NASA–ESSA, ESSA 1 Press Kit, ES 66–7, 30 Jan. 1966.

2. NASA, program office.

EXPLORER

1. A. Ruth Jarrell, Historical Office, MSFC, letter to Historical Staff, NASA, 16 Dec. 1963.

2. R. Cargill Hall, *Project Ranger: A Chronology,* JPL/HR-2 (Washington: NASA, 1971), p. 46; Eugene M. Emme, Historian, NASA, memorandum for the record (after conversation with Richard Hirsch, National Aeronautics and Space Council Staff), 26 Feb. 1970.

3. Robert W. Mulac, LaRC, letter to Historical Staff, NASA, 10 Dec. 1963.

4. NASA, News Release 70–203; NASA Headquarters Preliminary Results Press Briefing (transcript), 28 Dec. 1970; American Institute of Physics, News Release, 28 April 1971; John R. Holtz, Manager, Explorers, NASA, telephone interview, 20 March 1975.

5. *NASA Facts,* III, No. 4 (Washington: NASA, 1966); NASA, News Releases 65–333, 65–354, 68–16, 70–108; U.S. Congress, House of Representatives, Committee on Science and Astronautics' Subcommittee on Space Science and Applications, *Hearings . . . 1975 NASA Authorization,* Pt. 3, February and March 1974 (Washington: 1974), pp. 104–107, 189–190, 318.

6. John P. Donnelly, Assistant Administrator for Public Affairs, NASA, memorandum to Dr. John Naugle, Associate Administrator for Space Science, NASA, 8 June 1972; NASA program office; Henry L. Richter, Jr., ed., *Instruments and Spacecraft, October 1957–March 1965,* NASA SP-3028 (Washington: NASA, 1966), pp. 182, 236, 376, 522; and NASA, Historical Division, *Astronautics and Aeronautics: Chronology on Science, Technology, and Policy, 1968,* NASA SP–4010 (Washington: NASA, 1969), p. 182.

7. NASA, News Release 75–19; and European Space Research Organization, *Europe in Space,* March 1974, pp. 20–22.

8. Donnelly, memorandum to Naugle, 8 June 1972; and NASA, News Release 75–19.

9. James C. Fletcher, Administrator, NASA, "The NASA FY 1975 Budget," statement for 4 Feb. 1974 release; NASA, Fiscal Year 1975 Budget Briefing (transcript), 2 Feb. 1974; U.S. Congress, House of Representatives, Committee on Science and Astronautics' Subcom-

mittee on Space Science and Applications, *Hearings* . . . *1975 NASA Authorization*, Pt. 3, February and March 1974 (Washington: 1974), pp. 5–7, 259–268; and NASA, FY 1976 Budget Briefing (transcript), 1 Feb. 1975.

FR–1

1. NASA, News Release 63–49. The program was initiated by a Memorandum of Understanding signed by NASA and CNES 18 Feb. 1963.
2. John R. Holtz, Office of Space Science and Applications, NASA, telephone interview, 5 April 1967.
3. Homer E. Newell, Associate Administrator for Space Science and Applications, NASA, memorandum to NASA Headquarters and Field Centers, 24 May 1964. The memo was prepared for Dr. Newell's signature by FR–1 project manager John R. Holtz.

HEAO

1. NASA, Office of Technology Utilization, *A Long-Range Program in Space Astronomy*, NASA SP–213 (Washington: NASA, 1969), pp. 16–26.
2. Richard E. Halpern, Acting Director, Project HEAO, Office of Space Science and Applications, NASA, telephone interview, 14 May 1971.
3. NASA News Releases 73–40 and 74–240.

HEOS

1. ESRO, Communiqué No. 41, 8 March 1967; Ellen T. Rye, Office of International Affairs, NASA, telephone interview, 20 April 1967; and NASA, News Release 68–204.
2. ESRO, Communiqué No. 41; and Oscar E. Anderson, Director, International Organizations Division, Office of International Affairs, NASA, memorandum to Eugene M. Emme, Historian, NASA, 20 Dec. 1968.

INTASAT

1. GSFC, *Goddard News*, 20, No. 6 (September 1972), 1; NASA, News Releases 72–275 and 75–19; and program office.

INTELSAT

1. ComSat Corp., *Prospectus* (Washington: ComSat Corp., 1964), p. 14.
2. Larry Hastings, ComSat Corp., telephone interview, 21 April 1967; ComSat Corp., News Release 67–45; and NASA, program office.
3. Hastings, telephone interview; and ComSat Corp., News Releases 67–45 and 67–48. Press sources have erroneously referred to the satellite as "Lani Bird II."
4. Hastings, telephone interview; ComSat Corp., News Release 67–48; and NASA, program office.
5. ComSat Corp., News Releases 67–48 and 69–53; NASA, News Releases 68–195 and 69–6; and NASA, program office.
6. ComSat Corp., Public Relations Office, telephone interview, 22 Dec. 1971; and NASA, program office.

IRIS

1. NASA, program office.

ISIS

1. Ellen Thompson, Office of International Affairs, NASA, telephone interview, 31 July 1965.
2. U.S. Congress, Senate, Committee on Aeronautical and Space Sciences, *Hearings . . . NASA Authorization for FY 1972,* Pt. 1, March and April 1971 (Washington: 1971), p. 561.
3. NASA, News Release 71–72.
4. E. Sherrell Andrews, Office of International Affairs, NASA, telephone interview, 28 June 1971.

LAGEOS

1. Eberhard Rees, Director, MSFC, memorandum to Dale D. Myers, Associate Administrator for Manned Space Flight, NASA, 25 Jan. 1973.
2. *Ibid.;* U.S. Congress, House of Representatives, Committee on Science and Astronautics' Subcommittee on Space Science and Applications, *Hearings . . . 1975 NASA Authorization,* Pt. 3, February and March 1974 (Washington: 1974), pp. 104–106; NASA FY 1974 Budget News Conference (transcript), 23 Jan. 1973; and MSFC, News Releases 73–184 and 75–49.

NIMBUS

1. Robert F. Garbarini, Director of Applications, Office of Space Science and Applications, NASA, letter to Historial Staff, NASA, 30 Dec. 1963; and William K. Widger, Jr., *Meteorological Satellites* (New York: Holt, Rinehart and Winston, Inc., 1966), p. 153.

OAO

1. James E. Kupperian, Jr., Office of Space Flight Development, NASA, draft project outline, December 1958, with approval indicated by Gerhardt F. Schilling, Chief, Astronomy and Astrophysics Programs, Office of Space Flight Development, NASA; and Kupperian, letter to Historical Staff, NASA, 18 Nov. 1963.
2. NASA, "Proposed National Aeronautics and Space Administration Project" [first official OAO project document], 12 March 1959; and Kupperian, letter to Historical Staff, NASA, 18 Nov. 1963.
3. NASA, News Releases 72–141 and 72–156.

OFO

1. Robert W. Dunning, Office of Manned Space Flight [former OFO Experiment Program Manager, Office of Advanced Research and Technology], NASA, telephone interview, 14 May 1971.
2. NASA, program office.

3. *Ibid.*
4. Dunning, telephone interview, 14 May 1971; and NASA, program office.

OGO

1. Jack Posner, Office of Space Science and Applications, NASA, telephone interview, 10 Aug. 1965; and U.S. Congress, Senate, Committee on Aeronautical and Space Sciences, *Hearings . . . NASA Scientific and Technical Programs,* February and March 1961 (Washington: 1961), pp. 236–239.

OSO

1. Jack Posner, Office of Space Science and Applications, NASA, telephone interview, 10 Aug. 1965; and U.S. Congress, Senate, Committee on Aeronautical and Space Sciences, *Hearings . . . NASA Scientific and Technical Programs,* February and March 1961 (Washington: 1961), pp. 240–242.

PAGEOS

1. NASA, News Release 66–150.
2. Jack Posner, Office of Space Science and Applications, NASA, telephone interview, 10 Aug. 1965.
3. NASA, News Release 66–150.

PEGASUS

1. Raymond L. Bisplinghoff, Associate Administrator for Advanced Research and Technology, NASA, memorandum to Julian Scheer, Chairman, Project Designation Committee [and Assistant Administrator for Public Affairs], NASA, 23 Dec. 1963.
2. *Ibid.;* Milton B. Ames, Jr., Director, Space Vehicle Division, Office of Advanced Research and Technology, NASA, letter to Edward G. Uhl, President, Fairchild Stratos Corp., 21 July 1964; and NASA, News Release 64–203.

RELAY

1. Abe Silverstein, Director, Office of Space Flight Programs, NASA, memorandum to Robert C. Seamans, Jr., Associate Administrator, NASA, with approval signature of Dr. Seamans.
2. Robert Warren, Communication and Navigation Programs, Office of Space Science and Applications, NASA, letter to Historical Staff, NASA, 11 Dec. 1963.

SAN MARCO

1. Franco Fiorio, NASA Liaison for the Italian Space Commission, telephone interview, 2 Aug. 1965. Professor Broglio later became Chairman of the Italian Space Commission and San Marco Project General Director and Test Director; Professor Buongiorono, Assistant Project General Director; and Dr. Fiorio, NASA Liaison.
2. NASA, News Release 64–301; and NASA, Wallops Station, News Release 64–91.

SEASAT

1. Francis L. Williams, Director of Special Programs, Office of Applications, NASA, telephone interview, 9 June 1975; U.S. Congress, House of Representatives, Committee on Science and Astronautics' Subcommittee on Space Science and Applications, *Hearings . . . 1975 NASA Authorization,* Pt. 3, February and March 1974 (Washington: 1974), pp. 3–4, 270–271; and NASA, News Release 75–1.

SIRIO

1. NASA, News Release 70–42.
2. Barbara A. Goetz, Office of International Affairs, NASA, telephone interview, 7 July 1971.

SMS

1. NASA Ad Hoc Committee to Name Space Projects and Objects, minutes of meeting, 19 May 1960.
2. U.S. Congress, House of Representatives, Committee on Science and Astronautics, *Hearings . . . National Meteorological Satellite Program,* July 1961 (Washington: 1961), p. 32.
3. Michael L. Garbacz, Earth Observations Programs, Office of Space Science and Applications, NASA, telephone interview, 17 June 1971; NASA, News Release 63–18; and U.S. Congress. Senate, Committee on Aeronautical and Space Sciences, *Hearings . . . NASA Authorization for FY 1964,* Pt. 1, April 1963 (Washington: 1963), pp. 438–439, 441, 447.
4. NASA, Historical Office; and Garbacz, telephone interview, 17 June 1971.
5. Robert C. Seamans, Jr., Associate Administrator, NASA, letter to J. Robert Hollomon, Assistant Secretary for Science and Technology, Dept. of Commerce, 20 May 1965.
6. Garbacz, telephone interview, 11 June 1971; and NOAA, *Space: Environmental Vantage Point,* NOAA/PI 70033 (Washington: 1971), pp. 34–35.
7. NASA, News Releases 74–95, 74–154; NASA, program office.

SPHINX

1. NASA, News Release 74–25; NASA, Photo 74–H–92.

SYMPHONIE

1. NASA, News Release 74–316; European Space Research Agency, *Europe in Space,* March 1974, pp. 75–77; NASA, program office, 13 Dec. 1974; and Ed Arnone, *Today,* 27 Nov. 1974.
2. Jean-Pierre Pujes, Scientific Attaché, French Embassy, Washington, D.C., telephone interview, 9 May 1975; *Aviation Week & Space Technology,* 8 May 1967, p. 24; *Webster's New International Dictionary,* ed. 2 (unabridged).

SYNCOM

1. Alton E. Jones, GSFC, letter to Historical Staff, NASA, 7 April 1964; and GSFC, "Syncom Preliminary Project Development Plan," 5 Aug. 1961.

2. Robert C. Seamans, Jr., Associate Administrator, NASA, memorandum to Director, Office of Space Flight Programs, NASA, 17 Aug. 1961; and NASA, News Release 61–178.

TD

1. George D. Baker, Delta Project Office, GSFC, telephone interview, 23 July 1971.

2. NASA, News Release 66–332. TD was the second reimbursable launch under this agreement *(HEOS 1* was the first).

3. R. Lüst, "The European Space Research Organisation," *Science,* 149 (23 July 1965), 394–396. Negotiations for the purchase of U.S. launch vehicles were under way before the agreement was signed in 1966.

4. Baker, telephone interview; and "Europeans Reviewing Space Goals through Early 1970's," *Aviation Week & Space Technology,* 82 (14 June 1965), 200.

5. John L. Hess, "European Communication Satellite Seems Doomed," *New York Times,* 28 April 1968, p. 24; and Baker, telephone interview.

TELESAT

1. "Canadian Satellite," *Washington Post,* 16 April 1969, p. A17.

2. NASA, News Release 71–85.

3. *Ibid.*

TELSTAR

1. David Williamson, Jr., Office of Tracking and Data Acquisition, NASA, memorandum to Boyd C. Myers II (Chairman, NASA Project Designation Committee), Director, Program Review and Resources Management, NASA, 18 Oct. 1961; and Myers, memorandum to Robert C. Seamans, Jr., Associate Administrator, NASA, 30 Oct. 1961, with approval signature of Dr. Seamans, 2 Nov. 1961.

TIROS, TOS, AND ITOS

1. Robert F. Garbarini, Office of Space Science and Applications, NASA, letter to Historical Staff, NASA, 30 Dec. 1963.

2. NASA–ESSA, *ESSA 1* Press Kit, n.d.

3. *Ibid.*

4. NASA, News Release 66–115.

5. NASA, News Release 70–2.

6. NASA, program office.

VANGUARD

1. Milton W. Rosen, Office of Defense Affairs, NASA, letter to R. Cargill Hall, Lockheed Missiles & Space Co., 28 Aug. 1963.

2. *Ibid.;* Rosen, telephone interview, 16 Feb. 1965; and Chief of Naval Research, letter to Director, Naval Research Laboratory, 16 Sept. 1955.

WESTAR

1. Western Union Telegraph Co., *Communicator,* Summer 1973, pp. 4–5; UPI, "Western Union Proposes Satellite Telegram System," *New York Times,* 8 Nov. 1966, p. 15; NASA,

program office; and Dow Jones News Service, "Western Union Files Domestic Satellite Plan," Washington *Evening Star,* 31 July 1970, p. A14.

2. AP, "Satellite Relay for U.S. Approved," *New York Times,* 5 Jan. 1973, p. 1; "The Day of the Domsat," *Time,* 29 April 1974, p. 2; and NASA, program office.

3. Frances Shissler, Western Union Telegraph Co., McLean, Va., telephone interview, 2 April 1975.

PART III: SPACE PROBES

1. Milton Lehman, *This High Man: The Life of Robert H. Goddard* (New York: Farrar, Straus and Co., 1963), pp. 81–82.

2. William R. Corliss, *Space Probes and Planetary Exploration* (Princeton: Van Nostrand Co., 1965), p. 10.

3. Edgar M. Cortright, Assistant Director of Lunar and Planetary Programs, NASA, memorandum to NASA Ad Hoc Committee to Name Space Projects and Objects, 17 May 1960; and NASA, Ad Hoc Committee to Name Space Projects and Objects, minutes of meeting, 19 May 1960.

HELIOS

1. NASA, News Release 69–86. The memorandum of understanding was signed 10 June 1969.

2. Lloyd E. Jones, Jr., Office of International Affairs, NASA, telephone interview, 31 July 1969.

3. NASA , Historical Staff, *Astronautics and Aeronautics, 1963: Chronology on Science, Technology, and Policy,* NASA SP–4004 (Washington: NASA, 1964), p. 73; and *Astronautics and Aeronautics, 1965,* NASA SP–4006 (1966), p. 554.

LUNAR ORBITER

1. U.S. Congress, House of Representatives, Committee on Appropriations' Subcommittee on Independent Offices, *Hearings . . . FY 1966 Independent Offices Appropriations,* Pt. 2 (Washington: 1965), p. 858.

2. Jack Posner, Office of Space Science and Applications, NASA, telephone interview, 10 Aug. 1965.

3. NASA, News Release 68–23.

MARINER

1. Edgar M. Cortright, Assistant Director of Lunar and Planetary Programs, NASA, memorandum to NASA Ad Hoc Committee to Name Space Projects and Objects, 17 May 1960; and NASA Ad Hoc Committee to Name Space Projects and Objects, minutes of meeting 19 May 1960.

2. NASA, News Release 75–19.

PIONEER

1. David S. Akens, *Historical Origins of the George C. Marshall Space Flight Center,* MHM-1 (Huntsville: MSFC, 1960, p. 51, fn. 28; Eugene M. Emme, *Aeronautics and Astronautics: An American Chronology of Science and Technology in the Exploration of*

Space, 1915–1960 (Washington: NASA, 1961), pp. 102–103; and Emme, "Names of Launchings," enclosure to letter, 12 May 1960.

2. M/G Reginald M. Cram, Adjutant General, State of Vermont, letter to Stephen A. Saliga, Visual Aids Chief, NASA, 6 Feb. 1970; and Saliga, memorandum to Eugene M. Emme, Historian, NASA, 13 April 1972. M/G Cram was commander of the Air Force Orientation Group at the time.

3. John F. Clark, Director, GSFC, "Galactic/Jupiter Probes," address at the Fifth Goddard Memorial Symposium (AAS meeting), 14 March 1967; and Thomas P. Dallow, Office of Space Science and Applications, NASA, telephone interview, 14 June 1971.

4. U.S. Congress, House of Representatives, Committee on Science and Astronautics, *Hearings . . . 1969 NASA Authorization,* Pt. 3, February 1968 (Washington: 1968), pp. 207–208, 239–243.

5. George M. Low, Deputy Administrator, NASA, "Letter from Washington," *NASA Activities,* 5 (15 Dec. 1974), 3; and Peter W. Waller, Public Information Officer, ARC, telephone interview, 27 Feb. 1975. ARC Director Hans Mark had chosen this name from suggestions made by several persons in the Pioneer project and the Public Affairs Office.

6. NASA, News Release 75–19.

RANGER

1. Edgar M. Cortright, Assistant Director of Lunar and Planetary Programs, NASA, memorandum to NASA Ad Hoc Committee to Name Space Projects and Objects, 17 May 1960; and NASA Ad Hoc Committee to Name Space Projects and Objects, minutes of meeting, 19 May 1960.

2. Oran W. Nicks, Director of Lunar and Planetary Programs, NASA, in U.S. Congress, House of Representatives, Committee on Science and Astronautics, Subcommittee on NASA Oversight, *Hearings. . . Investigation of Project Ranger,* April 1964 (Washington: 1964), p. 56.

3. William H. Pickering, Director, JPL, letter to Abe Silverstein, Director of Space Flight Programs, NASA, 6 May 1960; and Muriel M. Hickey, Secretary to JPL Historian, letter to Historical Staff, NASA, 18 July 1967.

SURVEYOR

1. Edgar M. Cortright, Assistant Director of Lunar and Planetary Programs, NASA, memorandum to NASA Ad Hoc Committee to Name Space Projects and Objects, 17 May 1960; and NASA Ad Hoc Committee to Name Space Projects and Objects, minutes of meetings, 19 May 1960.

2. NASA, Office of Technology Utilization, *Surveyor Program Results,* NASA SP–184 (Washington: NASA, 1969), pp. v–vii.

VIKING

1. Walter Jakobowski, Office of Space Science and Applications, NASA, telephone interview, 16 July 1969.

2. Peter F. Korycinski, Office of the Director, LaRC, memorandum to Historical Division, NASA, 4 Sept. 1969.

PART IV: MANNED SPACE FLIGHT

1. Loyd S. Swenson, Jr., James M. Grimwood, and Charles C. Alexander, *This New Ocean: A History of Project Mercury,* NASA SP–4201 (Washington: NASA, 1966), p. 105.

2. John F. Kennedy [Special Message to the Congress, 25 May 1961], General Services Administration, National Archives and Records Service, Office of the Federal Register, *Public Papers of the Presidents of the United States: John F. Kennedy, 1961* (Washington: 1962), p. 404.

APOLLO

1. Merle G. Waugh, Office of Manned Space Flight, NASA, letter to James M. Grimwood, Historian, MSC, 5 Nov. 1963. The precedent of Mercury's name had been given consideration in NASA as early as 16 May 1960, when the Ad Hoc Committee to Name Space Projects and Objects "tentatively decided that the manned space flight programs will be named after the gods and heroes of mythology, thus continuing in the present class begun by 'Mercury.' " (NASA Ad Hoc Committee to Name Space Projects and Objects, minutes of meeting, 16 May 1960.)

2. Thomas Bulfinch, *Mythology,* Edmund Fuller, ed. (New York: Dell Publishing Co., Inc., 1959), pp. 17, 40ff.

3. Abe Silverstein, Director, Office of Space Flight Programs, NASA, memorandum to Harry J. Goett, Director, GSFC, July 25, 1960.

4. Julian W. Scheer, Assistant Administrator for Public Affairs, NASA, memorandum from Project Designation Committee, 9 June 1966.

5. George E. Mueller, Associate Administrator for Manned Space Flight, NASA, memorandum to Robert C. Seamans, Jr., Deputy Administrator, NASA, 9 Feb. 1967; Scheer, memorandum to Seamans, 17 Feb. 1967; Mueller, memorandum to Scheer, 28 March 1967; George M. Low, Deputy Director, MSC, letter to Mueller, 30 March 1967; Scheer, memorandum to distribution, 3 April 1967; and Mueller, TWX to KSC, MSFC, and MSC, Apollo and AAP Mission Designation, 24 March and 24 April 1967.

ASTP

1. *The Post-Apollo Space Program: Directions for the Future,* Space Task Group report to the President; General Services Administration, National Archives and Records Service, Office of the Federal Register, *Weekly Compilation of Presidential Documents,* 5 (22 Sept. 1969), 1291, and *Public Papers of the Presidents of the United States: Richard Nixon* (Washington: Government Printing Office, 1971), pp. 250–253; and Thomas O. Paine, "Man's Future in Space," 1972 Tizard Memorial Lecture, Westminster School, London, 14 March 1972.

2. Paine, "Man's Future in Space"; NASA, News Release, "Text of US/USSR Space Agreement," 24 May 1972, and News Release 72–109.

3. NASA, News Release, "Text," and News Release 72–109.

4. "Washington Roundup." *Aviation Week & Space Technology,* 96 (15 May 1972),13; Richard D. Lyons, "Chief Astronaut Foresees Further Cuts in the Corps," *New York Times,* 28 May 1972, 1; John P. Donnelly, Assistant Administrator for Public Affairs, NASA, memorandum to Dale D. Myers, Associate Administrator for Space Flight, NASA, 30 June

1972; and Apollo-Soyuz Test Project, Project Approval Document, Office of Manned Space Flight, NASA, 19 Dec. 1972 and 6 Oct. 1973.

5. NASA, News Release 75-9.

GEMINI

1. Glenn F. Bailey, MSC, interview, 13 Dec. 1966, reported by James M. Grimwood, Historian, MSC, 23 May 1968.

2. Alex P. Nagy, Office of Manned Space Flight, NASA, memorandum to George M. Low, Office of Manned Space Flight, NASA, 11 Dec. 1961; D. Brainerd Holmes, Director of Manned Space Flight Programs, NASA, memorandum to Associate Administrator, NASA, 16 Dec. 1961; Holmes, memorandum to Associate Administrator, NASA, 2 Jan. 1962; Robert C. Seamans, Jr., Secretary, USAF, letter to Eugene M. Emme, Historian, NASA, 3 June 1969; and desk calendar of Seamans, Associate Administrator, NASA, 15 Dec. 1961. Nagy's memorandum to Low proposing the name "Gemini" was dated 11 Dec., four days before Dr. Seamans' speech, but Dr. Seamans received his proposal and one from the member of the audience at about the same time.

3. Thomas Bulfinch, *Mythology,* Edmund Fuller, ed. (New York: Dell Publishing Co., Inc., 1959), pp. 130–131.

4. Holmes, memorandum to Associate Administrator, NASA, 16 Dec. 1961; and NASA, "NASA Photo Release 62–Gemini-2," 3 Jan. 1962.

5. Grimwood, *Project Mercury: A Chronology,* NASA SP–4001 (Washington: NASA, 1963), p. 133; Loyd S. Swenson, Jr., Grimwood, and Charles C. Alexander, *This New Ocean: A History of Project Mercury,* NASA SP–4201 (Washington: NASA, 1966), pp. 491–492; and Seamans, Associate Administrator, NASA, memorandum to George E. Mueller, Associate Administrator for Manned Space Flight, NASA, and Julian W. Scheer, Assistant Administrator for Public Affairs, NASA, 4 May 1965.

MERCURY

1. Thomas Bulfinch, *Mythology,* Edmund Fuller, ed. (New York: Dell Publishing Co., Inc., 1959), p. 18.

2. Loyd S. Swenson, Jr., James M. Grimwood, and Charles C. Alexander, *This New Ocean: A History of Project Mercury,* NASA SP–4201 (Washington: NASA, 1966), pp. 131–132.

3. *Ibid.,* p. 160. The earliest written record of the word "astronaut" is found in the writings of French poet Cyrano de Bergerac (1619–1655).

4. George M. Low, Chief, Manned Space Flight, Office of Space Flight Programs, NASA, memorandum to Abe Silverstein, Director, Office of Space Flight Programs, NASA, 12 Dec. 1958; and Swenson, Grimwood, and Alexander, *This New Ocean,* p. 342.

5. Swenson, Grimwood, and Alexander, *This New Ocean,* p. 342.

6. *Ibid.,* p. 368.

7. *Ibid.,* p. 446.

8. *Ibid.,* p. 470.

9. *Ibid.,* p. 492.

SKYLAB

1. John H. Disher, Deputy Director, Skylab Program, NASA, telephone interview, 19 Oct. 1971; NASA, Historical Staff, *Astronautics and Aeronautics, 1964: Chronology on*

Science, Technology, and Policy, NASA SP–4005 (Washington: NASA, 1965), pp. 145, 363; and *Astronautics and Aeronautics, 1965,* NASA SP–4006 (1966), p. 174.

2. Disher, telephone interview, 19 Oct. 1971; U.S. Senate, Committee on Aeronautical and Space Sciences, *Hearings . . . NASA Authorization for FY 1967,* February and March 1966 (Washington: 1966), pp. 163–166, 238–239; and NASA, Historical Staff, *Astronautics and Aeronautics, 1965,* pp. 418, 429.

3. NASA, News Releases 69–105 and 69–164.

4. L/C Donald L. Steelman, USAF, telephone interview, 12 Oct. 1971; and James A. Long, Assistant Executive Secretary, Office of the Administrator, NASA, notes of meeting of Project Designation Committee, 17 Feb. 1970.

SPACE SHUTTLE

1. LeRoy E. Day, Deputy Director of Space Shuttle Program, NASA, telephone interview, 2 April 1975; David Williamson, Jr., Assistant Administrator for Special Projects, NASA, telephone interview, April 1975; Ralph B. Jackson, Historical Monitor, Flight Research Center, telephone interview, 16 April 1975; and Norman L. Baker, Editor-in-Chief, *Defense/Space Business Daily,* telephone interview, 22 April 1975.

2. "Commercial Moon Flights Predicted within 30 Years," *Birmingham Post-Herald,* 9 Nov. 1962, report of Koelle's October paper for the American Rocket Society; and George S. James, "New Space Transportation Systems—An AIAA Assessment," unpublished draft paper [December 1972].

3. NASA, News Release, "Background Material and NASA Fiscal Year 1967 Budget Briefing," 22 Jan. 1966; and *The Space Program in the Post-Apollo Period,* report of President's Science Advisory Committee (Washington: Government Printing Office, February 1967), p. 37.

4. Wernher von Braun, "Crossing the Last Frontier," *Collier's,* 129, No. 12 (22 March 1952), 24–29, 72–74.

5. "Space Technology Highlights," *Astronautical Sciences Review,* October–December 1959, pp. 6–8, 29; and Courtney Sheldon, "Shuttle to Planets Awaits Development," *Christian Science Monitor,* 8 Dec. 1959.

6. Douglas Aircraft Co., "Summary Report and Rombus Systems Definition," Vol. 1, Douglas Report SM–42969, April 1963.

7. Walter J. Dornberger, "Space Shuttle for the Future: The Aerospaceplane," *Rendezvous,* 4, No. 1 (1965), 2–5; and Dornberger, "The Recoverable, Reusable Space Shuttle," *Astronautics & Aeronautics,* November 1965, pp. 88–94.

8. Robert F. Frietag, Director, MSF Field Center Development, NASA, "Space Shuttle Name," note for George E. Mueller, Associate Administrator for Manned Space Flight, NASA, 13 June 1969.

9. Mueller, briefing charts for address before British Interplanetary Society meeting, London, 10 Aug. 1968; and Mueller, prepared text, address before National Space Club, 26 Nov. 1968.

10. R. J. Wisniewski, Deputy Director for Programs, Programs and Resources Division, Office of Advanced Research and Technology, NASA, memorandum to OART Division Directors and Manager, SNPO, 10 Nov. 1969; Howard M. Weiss, Reliability and Quality Assurance Office (KR), NASA, memorandum to KR/Space Shuttle File, 19 Nov. 1969; NASA, News Release 69–70; NASA, Space Shuttle Conference Agenda, Washington, D.C., 16–17 Oct. 1969; and *The Post-Apollo Space Program: Directions for the Future,* Space Task Group report to the President, September 1969.

11. NASA, News Release 74–211; U.S. Congress, House of Representatives, Committee on Science and Astronautics' Subcommittee on Manned Space Flight, *Hearings . . . FY NASA Authorization*, Pt. 2, February and March 1974 (Washington: 1974), p. 9; and JSC, *Space Shuttle*, February 1975.

12. Joseph E. McGolrick, Manager for Advanced Programs and Technical Programs, Office of Space Science, NASA, telephone interview, 19 May 1975; McGolrick, Launch Vehicle Programs, NASA, memorandum for LV (Files), 5 Jan. 1961; and *The Post-Apollo Space Program: Directions for the Future.*

SPACELAB

1. NASA, News Releases 71–6 and 71–67; MSFC, News Releases 71–34 and 72–41; and U.S. Congress, House of Representatives, Committee on Science and Astronautics' Subcommittee on Manned Space Flight, *Hearings . . . 1973 NASA Authorization*, Pt. 2, February and March 1972 (Washington: 1972), pp. 238–245.

2. John P. Donnelly, Assistant Administrator for Public Affairs, NASA, memorandum to Dale D. Myers, Associate Administrator for Manned Space Flight, NASA, 8 June 1972.

3. NASA, News Release 73–191; and *Astronautics and Aeronautics: Chronology on Science, Technology, and Policy* for 1969, 1970, 1971, and 1972, NASA SP–4014 through SP–4017 (Washington: NASA, 1970, 1972, and 1974).

4. NASA, News Release 73–191; and JSC, *Space Shuttle* (Washington: GPO, February 1975).

5. James C. Fletcher, Administrator, NASA, memorandum to Administrators, Associate Administrators, Assistant Administrators, and Directors of Field Centers, 24 Sept. 1973.

6. Office of Manned Space Flight, NASA, Spacelab/CVT Program Approval Document (PAD), 4 Dec. 1974; NASA, News Releases 73–191 and 74–198; ESRO, News Release, 5 June 1974; ESRO, *Europe in Space*, March 1974, pp. 42–47; and Fletcher, prepared text for address before the National Space Club, Washington, D.C., 14 Feb. 1974.

7. NASA, News Release 75–28.

8. Douglas R. Lord, Director, Spacelab Program, NASA, prepared testimony for Hearings, FY 1976 NASA Authorization, before House Committee on Science and Technology's Subcommittee on Space Science and Applications, 20 Feb. 1975; NASA, News Release 73–191; and ESRO, News Releases, 4 Sept. 1974 and 21 Feb. 1975.

PART V: SOUNDING ROCKETS

1. See Homer E. Newell, ed., *Sounding Rockets* (New York: McGraw-Hill Book Co., Inc., 1959), pp. 1–2; and *Space: The New Frontier*, NASA EP–6 (Washington: NASA, 1966), pp. 37–41.

2. *Meteorological Satellites and Sounding Rockets*, NASA EP–27 (Washington: NASA [1965]), p. 17.

3. Robert H. Goddard, "A Method of Reaching Extreme Altitudes," *Smithsonian Miscellaneous Collections* (Washington: Smithsonian Institution, 1919), 71, No. 2, in Newell, "The Use of Rockets for Geophysical and Solar Research," Chap. II in *Sounding Rockets*, p. 28.

4. Eugene M. Emme, *Aeronautics and Astronautics: An American Chronology of Science and Technology in the Exploration of Space, 1915–1960* (Washington: NASA, 1961), p. 51; and Newell, "The Use of Rockets for Geophysical and Solar Research," Chap. II in *Sounding Rockets*, p. 28.

5. See Wallace W. Atwood, Jr., *The International Geophysical Year in Retrospect,* Dept. of State Publication 6850 (Washington: Dept. of State, 1959), from *Department of State Bulletin,* 11 May 1959.

6. GSFC, Sounding Rocket Division, *The United States Sounding Rocket Program,* X-740-71-337 preprint (Greenbelt, Md.: GSFC, July 1971), p. 1; and NASA, News Release 75-19.

7. GSFC, *The United States Sounding Rocket Program,* pp. 2-3, 36.

8. *Ibid.,* p. 38; John R. Holtz, Manager, Explorers, Sounding Rockets and Balloons, Office of Space Science, NASA, telephone interview, 16 Dec. 1974.

AEROBEE

1. Peter T. Eaton, Office of Space Science and Applications, NASA, letter to Historical Staff, NASA, 2 May 1967; and James A. Van Allen, Eleanor Pressly, and John W. Townsend, Jr., "The Aerobee Rocket," Chap. IV in Homer E. Newell, ed., *Sounding Rockets* (New York: McGraw-Hill Book Co., Inc., 1959), p. 57.

2. John W. Townsend, Jr., Eleanor Pressly, Robert M. Slavin, and Louis Kraff, Jr., "The Aerobee-Hi Rocket," Chap. V in Newell, *Sounding Rockets,* p. 71.

3. GSFC, *Encyclopedia: Satellites and Sounding Rockets of Goddard Space Flight Center, 1959-1969* (Greenbelt, Md.: GSFC [1970]), p. 321; William R. Corliss, *NASA Sounding Rockets, 1958-1968: A Historical Summary,* NASA SP-4401 (Washington: NASA, 1971), pp. 79-80; and Herbert J. Honecker, Advanced Vehicles Section, Flight Performance Branch, Sounding Rocket Division, GSFC, memorandum to John H. Lane, Head, Flight Performance Branch, 10 Jan. 1975.

4. GSFC, Sounding Rocket Division, *The United States Sounding Rocket Program,* X-740-71-337 preprint (Greenbelt, Md.: GSFC, July 1971), pp. 38, 47; and Edward E. Mayo, Flight Performance Branch, Sounding Rocket Division, GSFC, information sent to Historical Office, NASA, 30 Jan. 1975.

APACHE

1. Alfred Rosenthal, *Venture into Space: Early Years of Goddard Space Flight Center,* NASA SP-4301 (Washington: NASA, 1968), pp. 127-129; *Space: The New Frontier,* NASA EP-6 (Washington: NASA, 1966), p. 38; Vehicles Section, Spacecraft Integration and Sounding Rocket Division, GSFC, telephone interview, 19 March 1970; and GSFC, Sounding Rocket Division, *The United States Sounding Rocket Program,* X-740-71-337 preprint (Greenbelt, Md.: GSFC, July 1971), pp. 38, 47.

2. R. Gilbert Moore and John M. McGarry, Astro-Met Plant, Thiokol Corp., joint telephone interview, 21 May 1975; and John H. Lane, Head, Flight Performance Branch, Sounding Rocket Division, GSFC, information sent to Historical Office, 30 Jan. 1975.

ARCAS

1. Atlantic Research Corp., Announcement released by U.S. Army Missile Support Agency, 26 Jan. 1959.

2. Peter T. Eaton, Office of Space Science and Applications, NASA, letter to Historical Staff, NASA, 2 May 1967.

3. W. C. Roberts, Jr., and R. C. Webster, Atlantic Research Corp., "Arcas Rocketsonde System Development," 3 Sept. 1959; and GSFC, *Encyclopedia: Satellites and Sounding*

Rockets of Goddard Space Flight Center, 1959–1969 (Greenbelt, Md.: GSFC [1970]), p. 321.

4. John H. Lane, Head, Flight Performance Branch, Sounding Rocket Division, GSFC, information sent to Historical Office, NASA, 30 Jan. 1975.

ARGO

1. Peter T. Eaton, Office of Space Science and Applications, NASA, letter to Historical Staff, NASA, 2 May 1967.

2. Alfred Rosenthal, *Venture into Space: Early Years of Goddard Space Flight Center,* NASA SP–4301 (Washington: NASA, 1968), pp. 127–129.

3. John H. Lane, Head, Flight Performance Branch, Sounding Rocket Division, GSFC, information sent to Historical Office, NASA, 30 Jan. 1975; Eaton, letter; and Wallops Station, News Release 71–12.

ARIES

1. U.S. Congress, House of Representatives, Committee on Science and Astronautics' Subcommittee on Space Science and Applications, *Hearings. . . 1975 NASA Authorization,* Pt. 3, February and March 1974 (Washington: 1974), pp. 117, 456, 560–561; John R. Holtz, Manager, Explorers, Sounding Rockets and Balloons, Office of Space Science, NASA, telephone interview, 16 Dec. 1974; and John H. Lane, Head, Flight Performance Branch, Sounding Rocket Division, GSFC, information sent to Historical Office, 30 Jan. 1975.

2. Holtz, telephone interview.

3. *Hearings. . . 1975 NASA Authorization,* Pt. 3, pp. 116–117.

4. William H. Conway, Space Science Division, Naval Research Laboratory, telephone interview, 16 Dec. 1974; and Herbert J. Honecker, Advanced Vehicles Section, Flight Performance Branch, Sounding Rocket Division, GSFC, memorandum to John H. Lane, Head, Flight Performance Branch, 10 Jan. 1975.

ASP

1. Robert B. Cox, "Asp," in Homer E. Newell, ed., *Sounding Rockets* (New York: McGraw-Hill Book Co., Inc., 1959), p. 105.

2. Peter T. Eaton, Office of Space Science and Applications, NASA, letter to Historical Staff, NASA, 2 May 1967.

BLACK BRANT

1. A. W. Fia, Vice President, Rocket and Space Division, Bristol Aerospace Ltd., "Canadian Sounding Rockets: Their History and Future Prospects," *Canadian Aeronautics and Space Journal,* 20, No. 8 (October 1974), 396–406.

2. Fia, letter to Historical Office, NASA, 27 Dec. 1974.

3. Fia, "Canadian Sounding Rockets"; and NASA, sounding rocket launch reports, 1973, 1974.

4. Fia, "Canadian Sounding Rockets"; and Richard H. Ott, Jr., Flight Performance Branch, Sounding Rocket Division, GSFC, information sent to Historical Office, NASA, 30 Jan. 1975.

CAJUN

1. William J. O'Sullivan, Jr., "Deacon and Cajun," Chap. VI in Homer E. Newell, ed., *Sounding Rockets* (New York: McGraw-Hill Book Co., Inc., 1959), pp. 100–101; Peter T. Eaton, Office of Space Science and Applications, NASA, letter to Historical Staff, NASA, 2 May 1967; and R. Gilbert Moore and John M. McGarry, Astro-Met Plant, Thiokol Corp., joint telephone interview, 21 May 1975.

2. O'Sullivan, "Deacon and Cajun," pp. 96–97, 100–101.

3. Alfred Rosenthal, *Venture into Space: Early Years of Goddard Space Flight Center,* NASA SP–4301 (Washington: NASA, 1968), pp. 127–129; and GSFC, Sounding Rocket Division, *The United States Sounding Rocket Program,* X–740–71–337 preprint (Greenbelt, Md.: GSFC, July 1971), p. 38.

HAWK

1. Bobby J. Flowers, Hawk project engineer, Wallops Station, telephone interview, 7 Jan. 1975.

2. *Ibid.;* and John H. Lane, Head, Flight Performance Branch, Sounding Rocket Division, GSFC, telephone interview, 17 Dec. 1974.

MALEMUTE

1. William J. Bolster, Flight Performance Branch, Sounding Rocket Division, GSFC, information sent to Historical Office, NASA, 30 Jan. 1975; and R. Gilbert Moore and John M. McGarry, Astro-Met Plant, Thiokol Corp., joint telephone interview, 21 May 1975.

2. Bolster, information sent Historical Office.

NIKE

1. Eugene M. Emme, *Aeronautics and Astronautics: An American Chronology of Science and Technology in the Exploration of Space, 1915–1960* (Washington: NASA, 1961), p. 49; and Peter T. Eaton, Office of Space Science and Applications, NASA, letter to Historical Staff, NASA, 2 May 1967.

2. Edward E. Mayo, Flight Performance Branch, Sounding Rocket Division, GSFC, information sent to Historical Office, NASA, 30 Jan. 1975.

TERRIER

1. Edward E. Mayo, Flight Performance Branch, Sounding Rocket Division, GSFC, information sent to Historical Office, NASA, 30 Jan. 1975.

TOMAHAWK

1. R. Gilbert Moore and John M. McGarry, Astro-Met Plant, Thiokol Corp., joint telephone interview, 21 May 1975.

2. GSFC, Sounding Rocket Branch, *The United States Sounding Rocket Program,* X–740–71–337 preprint (Greenbelt, Md.: GSFC, July 1971), pp. 38, 47.

PART VI: INSTALLATIONS

AMES

1. Jerome C. Hunsaker, *Forty Years of Aeronautical Research,* Smithsonian Publication 4237, reprint from *Smithsonian Report for 1955* (Washington: Smithsonian Institution, 1956), p. 261; and Robert L. Rosholt, *An Administrative History of NASA, 1958–1963,* NASA SP–4101 (Washington: NASA, 1966), p. 21.

2. Hunsaker, *Forty Years,* pp. 255, 265–266.

3. Eugene M. Emme, *Aeronautics and Astronautics: An American Chronology of Science and Technology in the Exploration of Space, 1915–1960* (Washington: NASA, 1961), p. 102.

ERC

1. NASA, News Release 64–219; NASA, Announcement 64–189; and NASA, Circulars 320 and 321.

2. NASA, News Release 69–171; General Services Administration, National Archives and Records Service, Office of the Federal Register, *Weekly Compilation of Presidential Documents,* 6, No. 13 (30 March 1970), 446; and Department of Transportation, Release 6870.

FRC

1. Eugene M. Emme, *Aeronautics and Astronautics: An American Chronology of Science and Technology in the Exploration of Space, 1915–1960* (Washington: NASA, 1961), p. 55.

2. *Forty-second Annual Report of the National Advisory Committee for Aeronautics, 1956* (Washington: NACA, 1957), pp. 6–9; and "A Brief History of the NASA Flight Research Center," enclosure in Jack Fischel, FRC, letter to Manley Hood, Office of the Director, ARC, 2 Feb. 1968.

3. Emme, *Aeronautics and Astronautics, 1915–1960,* p. 113.

GSFC

1. Public Law 85–657, 14 Aug. 1958, in Robert L. Rosholt, *An Administrative History of NASA, 1958–1963,* NASA SP–4101 (Washington: NASA, 1966), p. 79.

2. President Dwight D. Eisenhower, Executive Order No. 10783, 1 Oct. 1958; and NASA, Beltsville Space Center, General Notice No. 1, 15 Jan. 1959, in Alfred Rosenthal, *Venture into Space: Early Years of Goddard Space Flight Center,* NASA SP–4301 (Washington: NASA, 1968), Appendix H.

3. Rosenthal, *Venture into Space,* p. 29.

4. See Milton Lehman, *This High Man: The Life of Robert H. Goddard* (New York: Farrar, Straus and Co., 1963).

5. Rosholt, *Administrative History,* p. 124; NASA, News Release 61–15; and Rosenthal, *Venture into Space,* pp. 59–60. NASA announced plans 2 Jan. 1961 for establishment of the New York office and formal operations began in May of that year.

JPL

1. Eugene M. Emme, *Aeronautics and Astronautics: An American Chronology of Science and Technology in the Exploration of Space, 1915–1960* (Washington: NASA, 1961), p. 34.

2. *Ibid.*, p. 48.

3. President Dwight D. Eisenhower, Executive Order No. 10793, 3 Dec. 1958, in Robert L. Rosholt, *An Administrative History of NASA, 1958–1963*, NASA SP–4101 (Washington: NASA, 1966), p. 47.

JSC

1. Robert L. Rosholt, *An Administrative History of NASA, 1958–1963*, NASA SP–4101 (Washington: NASA, 1966), pp. 124, 214.

2. Loyd S. Swenson, Jr., James M. Grimwood, and Charles C. Alexander, *This New Ocean: A History of Project Mercury*, NASA SP–4201 (Washington: NASA, 1966), p. 392.

3. *Congressional Record—Senate*, 26 Jan. 1973 (Washington: 1973), p. S1344; *Congressional Record—House*, 29 Jan. 1973, p. H553; 7 Feb. 1973, p. H77.

4. Christopher C. Kraft, Jr., Director, Manned Spacecraft Center, letter to Sen. Frank E. Moss, Chairman, Senate Committee on Aeronautical and Space Sciences, 1 Feb. 1973; James C. Fletcher, Administrator, NASA, letter to Rep. Olin E. Teague, Chairman, House Committee on Science and Astronautics, 5 Feb. 1973; Gerald J. Mossinghoff, Assistant General Counsel for General Law, NASA, telephone interview, 28 June 1974; *Congressional Record—Daily Digest*, 2 Feb. 1973, p. D71.

5. *Congressional Record—Senate*, 6 Feb. 1973, pp. S2229–2230; *Congressional Record—House*, 7 Feb. 1973, pp. H838–839; *Congressional Record—Daily Digest*, 20 Feb. 1973, p. D117; NASA Notice 1132, 17 Feb. 1973.

6. U.S. Congress, Senate, Committee on Aeronautics and Space Sciences, *Tenth Anniversary, 1958–1968*, S. Doc. 116, 19 July 1968 (Washington: 1968), and *Statements by Presidents of the United States on International Cooperation in Space*, S. Doc. 92–40, 24 Sept. 1971 (Washington: 1971), pp. 55–90; and JSC, transcript, "Dedication of Lyndon B. Johnson Space Center," 27 Aug. 1973.

7. White House Release, Key Biscayne, Florida, 19 [17] Feb. 1973.

8. MSC, Weekly Activities Report, 24–30 June 1962; James M. Grimwood, Historian, MSC; MSC, Message 7–02, 2 July 1962.

9. MSC, Announcement 65–86, 25 June 1965; and Grimwood.

KSC

1. Frank E. Jarrett, Jr., and Robert A. Lindemann, *Historical Origins of NASA's Launch Operations Center to July 1, 1962*, KHM–1 (Cocoa Beach: KSC, 1964), pp. 21–22, 32. The Long Range Proving Ground was operated by the Air Force and activated 1 Oct. 1949, at what had been the Banana River Naval Air Station.

2. *Ibid.*, p. 68; and Robert L. Rosholt, *An Administrative History of NASA, 1958–1963*, NASA SP–4101 (Washington: NASA, 1966), p. 123.

3. Jarrett and Lindemann, *Origins of LOC*, p. 54.

4. NASA, Circular 208, in Jarrett and Lindemann, *Origins of LOC*, Appendix A.

5. President Lyndon B. Johnson, Executive Order No. 11129, 29 Nov. 1963, in Angela C. Gresser, *Historical Aspects Concerning the Redesignation of Facilities at Cape Canaveral*, KHN–1 (Cocoa Beach: KSC, 1964), p. 15.

6. Rosholt, *Administrative History*, pp. 214–215.

7. Gresser, *Redesignation of Facilities*, p. 9.

8. KSC, Announcement, 26 July 1965.

9. NASA, Circular 267–A.

10. Rosholt, *Administrative History,* p. 124; and NASA, Circular 208, in Jarrett and Lindemann, *Origins of LOC,* Appendix A.

11. KSC Organization Chart, in NASA Management Instruction 1142.2, 30 Sept. 1965; and NASA, *Weekly Bulletin,* 1–65, 15 Nov. 1964, pp. 1–2.

LaRC

1. Jerome C. Hunsaker, *Forty Years of Aeronautical Research,* Smithsonian Publication 4237, reprint from *Smithsonian Report for 1955* (Washington: Smithsonian Institution, 1956), pp. 250–251. John F. Victory's Day Book states that ground was broken at Langley Field 17 July 1917.

2. Bessie Zaban Jones, *Lighthouse of the Skies: The Smithsonian Astrophysical Observatory: Background and History, 1846–1955,* Smithsonian Publication 4612 (Washington: Smithsonian Institution, 1965), pp. 105, 155–158.

3. Hunsaker, *Forty Years,* p. 251.

4. Eugene M. Emme, *Aeronautics and Astronautics: An American Chronology of Science and Technology in the Exploration of Space, 1915–1960* (Washington: NASA, 1961), p. 102.

LeRC

1. Jerome C. Hunsaker, *Forty Years of Aeronautical Research,* Smithsonian Publication 4237, reprint from *Smithsonian Report for 1955* (Washington: Smithsonian Institution, 1956); and Robert L. Rosholt, *An Administrative History of NASA, 1958–1963,* NASA SP–4101 (Washington: NASA, 1966), p. 21.

2. Michael Keller, author of "Thirty-Year History of Langley Research Center," unpublished, interview, 24 Aug. 1966.

3. V/A Emory S. Land, "George William Lewis: An Address," in *George William Lewis Commemoration Ceremony* [program] (Cleveland: NACA Lewis Flight Propulsion Laboratory, 28 Sept. 1948).

4. Hunsaker, *Forty Years,* plate 2 (caption), and p. 242.

5. Eugene M. Emme, *Aeronautics and Astronautics: An American Chronology of Science and Technology in the Exploration of Space, 1915–1960* (Washington: NASA, 1961), p. 102.

6. Hugh W. Harris, LaRC, letter to Historical Staff, NASA, 2 May 1967; and Lewis Research Center, *Plum Brook Station* (fact sheet), March 1967.

7. NASA, News Release 73-3; LeRC, News Release 73-10a; Hugh W. Harris, Public Information Office, LeRC, telephone interview, 24 June 1974; U.S. Congress, House of Representatives, *Authorizing Appropriations to the National Aeronautics and Space Administration,* Rpt. No. 93-983, 10 April 1974 (Washington: 1974); and Public Information Office, LeRC, "Response to Queries" [June 1974].

MSFC

1. David S. Akens, *Historical Origins of the George C. Marshall Space Flight Center,* MHM–1 (Huntsville: MSFC, 1960); and Frank E. Jarrett, Jr., and Robert A. Lindemann, *Historical Origins of NASA's Launch Operations Center to July 1, 1962,* KHM–1 (Cocoa Beach: KSC, 1964), p. x.

2. Robert L. Rosholt, *An Administrative History of NASA, 1958–1963,* NASA SP–4101 (Washington: NASA, 1966), p. 120.

3. President Dwight D. Eisenhower, Executive Order 10870, 17 March 1960, in Akens, *Origins of MSFC,* p. 77; and Rosholt, *Administrative History,* p. 120.

4. Akens, *Origins of MSFC,* p. 81; cf. Rosholt, *Administrative History,* pp. 118–123.

5. NASA, News Release 61–201; and Robert C. Seamans, Jr., Associate Administrator, NASA, letter to MSFC, in MSFC, Historical Office, *History of the George C. Marshall Space Flight Center: July 1–December 31, 1961,* MHM-4 (Huntsville: MSFC, 1962), I, 38.

6. MSFC, News Release 65–167. The new name was instituted in a new organization chart signed by NASA Administrator James E. Webb 28 May 1965, but not issued as of 1 July 1965.

7. NASA, News Release 61–236; Robert P. Young, Executive Assistant, Office of the Administrator, NASA, letter to Army Chief of Engineers, 29 Nov. 1961; NASA, Circular 188; Seamans, letter to MSFC, in MSFC, Historical Office, *History of MSFC,* p. 38; and L. L. Jones, Historical Office, MSFC, letter to Historical Staff, NASA, 26 Aug. 1966.

8. MSFC, News Release 65–167.

9. NASA, News Release 70–98; and MSFC, *Marshall Star,* 10 Nov. 1970, pp. 1, 4.

10. NASA, News Release 70–147; Leonard Jaffe, Associate Administrator for Space Science and Applications, memorandum to Deputy Associate Administrator for Space Science and Applications (Applications), NASA, 6 May 1970; Jaffe, memorandum for the record, "Use of the Mississippi Test Facility (MTF)," 11 June 1970; S. M. Smolensky, Special Assistant for Policy, NASA, to J. Cramer, Legislative Affairs Officer (OMSF), NASA, 7 April 1971; and NASA, News Release 74–159.

11. NASA, News Releases 71–30 and 72–167.

12. NASA, News Release 74–159.

NSTL

1. NASA, News Release 74–159.

2. NASA, News Release 70–98; and MSFC, *Marshall Star,* 10 Nov. 1970, pp. 1, 4.

3. NASA, News Releases 70–98, 70–114, 70–141, and 70–147.

4. NASA, News Releases 71–30 and 72–167.

5. NASA, News Releases 74–159 and 70–147.

6. NASA, News Release 74–159.

WALLOPS

1. Joseph A. Shortal, "History of Wallops Station," Part I: "Origin and Activities through 1949" (comment ed.), July 1967, pp. 47, 49, 50.

2. Robert L. Rosholt, *An Administrative History of NASA, 1958–1963,* NASA SP–4101 (Washington: NASA, 1966), pp. 48, 81, Figs. 3–1, and Appendix B.

3. NASA, News Release, 24 Jan. 1959.

4. LaRC, *Air Scoop,* 22 July 1960. The *Air Scoop* credits the Accomac, Va., *Peninsula Enterprise* newspaper for this information.

5. NASA, Special Announcement, 25 April 1974; Wallops Station News Release 74–2.

PHOTOGRAPHS OF MYTHOLOGICAL FIGURES

Atlas, sculpture in the National Museum, Naples. Cosmos Pictures Co., New York, from the Library of Congress Prints and Photographs Department.

Centaur. Reproduced by the Library of Congress from Richard P. Knight, *The Symbolical Language of Ancient Art and Mythology,* Alexander Wilder trans. (New York: J. W. Bouton, 1892).

Juno, in the Uffizi Gallery, Florence. From the Library of Congress Prints and Photographs Department.

Saturn. Reproduced by the Library of Congress from Bernard de Montfaucon, *L'Antiquité Expliqué et Representée en Figures,* Vol. 1, *Lex Dieux des Grecs et des Romaines,* ed. 2 (Paris: 1722).

Thor, statue by B. E. Fogelberg. Reproduced by the Library of Congress from Donald A. MacKenzie, *Teutonic Myth and Legend* (New York: William H. Wise & Co., 1934).

Ocean, a Titan. Reproduced from de Montfaucon.

Apollo, fourth century sculpture in the Vatican Museum, Rome. Photograph from George Washington University Department of Art.

Castor and Pollux, the Gemini, in the Museum of Fine Arts, Boston, Mass. Reproduced from the collections of the Library of Congress.

Mercury, bronze by Giovanni Bologna in the Mellon Collection, National Gallery of Art, Washington, D.C. Photograph from the National Gallery.

Winged Nike (Victory of Samothrace) in the Louvre, Paris. The Perry Pictures, Boston ed., from the Library of Congress Prints and Photographs Department.

INDEX

AAP (Apollo Applications Program), 109

Able (launch vehicle upper stage), 5 il., 9 il., 12

ABMA. See Army Ballistic Missile Agency.

Ad Hoc Committee to Name Space Projects and Objects, 67 n., 183, 199

Advanced Mercury (Gemini program), 104

Advanced Orbiting Solar Observatory (AOSO), 84

Advanced Planetary Probe, 90

Advanced Research Projects Agency (ARPA), 5, 6, 11, 12, 15, 17, 76, 88

Advanced Saturn (Saturn C-5; renamed Saturn V), 20

Advanced Synchronous Orbit Satellite, 36

Advanced Synchronous Satellite, 36

Advanced Syncom (synchronous communications satellite), 36

Advanced Technological Satellite, 36

AE (Atmosphere Explorer, scientific satellite), 51 il.

Aeolus (mythological Greek god), 41

Aerobee (sounding rocket), 122–124
    Aerobee 150 and 150A, 122 il.–123
    Aerobee 170 and 170A, 123, 133
    Aerobee 200 and 200A, 123, 133
    Aerobee 350, 122 il., 124, 127, 133
    Aerobee-Hi, 123

Aerobee (Vanguard launch vehicle stage), 79

Aerojet Engineering Corp. (later Aerojet-General Corp.), 123

Aerolab Co. (later division of Atlantic Research Corp.), 125

Aeronaut (balloon pioneers), 107

Aeronautical Satellite (AEROSAT), 47

Aeronomy Explorer (scientific satellite), 49

Aeros (GRS–A-2; aeronomy satellite), 33 il.
    Aeros 2 (Aeros-B), 33, 38 n.

Aeros (mythological Greek god), 33, 69

Aeros (Synchronous Meteorological Satellite project), 33, 69–71, 79

AEROSAT (Aeronautical Satellite), 47

Aerospaceplane, 111–114

AES (Apollo Extension System), 112

AFBMD (Air Force Ballistic Missile Division), 94

AFCRL (Air Force Cambridge Research Laboratories), 133

Agena (Beta Centauri; star), 6–7

Agena (launch vehicle upper stage), 6 il.–7 il., 9 il., 22, 152
    Agena A, 7
    Agena B, 7
    Agena D, 7

Agreement Concerning Cooperation in the Exploration and Use of Outer Space for Peaceful Purposes, 103

Agriculture, Department of, 42–45, 142

Air Density Satellite (Explorer), 48

Air Force Ballistic Missile Division (AFBMD), 88–89

Air Force Cambridge Research Laboratories (AFCRL), 133

Air Force One (aircraft), 103

Air Research and Development Command (ARDC), 22, 25

Aircraft Engine Research Laboratory, NACA (renamed Lewis Flight Propulsion Laboratory Sept. 1948), 152

Aldrin, Edwin E., Jr., 98 il.

Algol (Scout first stage), 20

Algol (star), 20

All-purpose Rocket for Collecting Atmospheric Soundings (sounding rocket). See Arcas.

Allegany Ballistics Laboratory, 131, 188

Alouette (topside sounder satellite), 7, 35
    Alouette 1, 34 il., 35, 58
    Alouette 2, 35, 58

Altair (Scout fourth stage), 21

Altair (star), 21

AMB (Astronomy Missions Board, NASA), 52

America (CSM-114, on Apollo 17), 100

American Indian, 125, 131, 133

American Telephone & Telegraph Co. (AT&T), 75
    Bell Telephone Laboratories, Inc., 40, 75

Ames, Joseph S., 138 il., 139

Ames Aeronautical Laboratory, NACA (became Ames Research Laboratory, NASA, Oct. 1958), 139
Ames Research Center (ARC), 62, 137, 138 il., 139
AMPS (Atmospheric, Magnetospheric, and Plasmas-in-Space) laboratory, 118
AMR (Atlantic Missile Range, USAF; formerly Long Range Proving Ground; now Eastern Test Range), 149–150
AMROO (Atlantic Missile Range Operations Office, MSFC), 149
AMSAT OSCAR 7 (Orbiting Satellite Carrying Amateur Radio), 15, 55,
Anderson, O. E., 58
Anik (Canadian Telesat telecommunications satellite), 75
   Anik 1 (Telesat-A), 75
   Anik 2 (Telesat-B), 75
   Anik 3 (Telesat-C), 75
ANS (Astronomical Netherlands Satellite; also Netherlands Astronomical Satellite), 34 il.
ANS Program Authority, 35
Antares (LM-8, on *Apollo 14*), 100
Antares (Scout third stage), 20
Antares (star), 20
AOSO (Advanced Orbiting Solar Observatory), 84
Apache (American Indian tribe), 125
Apache (sounding rocket upper stage), 124 il.–125, 138
APL (Applied Physics Laboratory), 126
Apollo (manned space program; Project Apollo), 62, 97, 98–101, 109, 113
   launch vehicle, 3, 18–20, 58, 155–158
   planning for, 85, 91, 104
   program names, ix–x, 99–101
   Center responsibility, 147, 149, 155–158
Apollo (mythological Greek god), 98 il., 99
Apollo (spacecraft), 16, 19, 98 il.–101, 102 il.–104, 111
   Apollo 1 (AS–204), 100
   Apollo 4 (AS–501), 100–101
   Apollo 7 (AS–205), 19
   Apollo 8 (AS–503), 18 il., 20, 101
     photo by, 96 il., 146 il.
   Apollo 9 (AS–504), 100
   Apollo 10 (AS–505), 100
   Apollo 11 (AS–506), 98 il., 100, 101, 103, 158
   Apollo 12 (AS–507), x, 100
   Apollo 13 (AS–508), 100
   Apollo 14 (AS–509), 100
   Apollo 15 (AS–510), 100

Apollo 15 Subsatellite, x
Apollo 16 (AS–511), 100
Apollo 17 (AS–512), 20, 100, 101
Apollo Applications Program (AAP), 109
Apollo Extension System (AES), 109
Apollo Saturn (AS; Apollo space vehicle and flight), x
Apollo Saturn IB
   AS–201, x, 100
   AS–202, 100
   AS–203, 100
   AS–204 (Apollo 1), 100
   AS–205 (*Apollo 7*), 19
   AS–210 (ASTP), 19, 97, 101–104, 147, 155
Apollo Saturn V
   AS–501 (*Apollo 4*), 100–101
   AS–503 (*Apollo 8*), 18 il., 20, 96, 98, 101, 146
   AS–504 (*Apollo 9*), 100
   AS–505 (*Apollo 10*), 100
   AS–506 (*Apollo 11*), 98 il., 100, 101, 103, 158
   AS–507 (*Apollo 12*), x, 100
   AS–508 (*Apollo 13*), 100
   AS–509 (*Apollo 14*), 100
   AS–510 (*Apollo 15*), x, 100
   AS–511 (*Apollo 16*), 100
   AS–512 (*Apollo 17*), 20, 100, 101
Apollo-Soyuz Test Project (ASTP), 19, 97, 101–102 il.–104, 147, 155
Apollo Telescope Mount (ATM, on *Skylab 1*), 109–110 il., 155
Apollo X (Extended Apollo), 109
Applications Explorer (satellite), 52
Applications Technology Satellite (ATS), ix, 12, 25, 30, 36–37 il., 71
Applied Physics Laboratory (APL), 123
Aquarius (LM-7, on *Apollo 13*), 100
Aquila (constellation), 21
ARC. See Ames Research Center.
Arcas (sounding rocket), 123, 125
   Boosted Arcas, 125
   Boosted Arcas II, 125
   Super Arcas, 125
Arcon (sounding rocket), 125
ARDC (Air Research and Development Command), 22, 25
Argo (mythological Greek ship), 125
Argo (sounding rocket), 125–127
   Argo D–4 (Javelin), 123, 126 il.–127
   Argo D–8 (Journeyman), 126 il.–127
Argonaut (Greek mythology), 107
Argus (builder of *Argo,* mythology), 125 n.

INDEX

Ariel (ionospheric research satellite), 15, 35–36
  Ariel 1 (UK-1), 35
  Ariel 2 (UK-2), 36
  Ariel 3 (UK-3), 36
  Ariel 4 (UK-4), 34 il., 36
  Ariel 5 (UK-5), 36
Ariel (spirit of the air), 35
Aries (constellation), 127
Aries (sounding rocket), 123, 127
Army Ballistic Missile Agency (ABMA), 15, 17,
  49, 88–89, 144, 154–155
  Development Operations Division (transferred
  to MSFC March 1960), 17, 154–155
ARPA. See Advanced Research Projects
  Agency.
Arritt, Robert D., 127
AS. See Apollo Saturn.
Asp (sounding rocket), 127, 128 il.
Asteroid Belt, 90
ASTP (Apollo-Soyuz Test Project), 19, 97,
  101–102 il.–104, 147, 155
Astrobee (sounding rocket), 122–124
  Astrobee 1500, 122 il., 124
  Astrobee F, 124
Astronaut, derivation of term, 107
Astronaut, Project (proposed name), 107
Astronomical Netherlands Satellite (*ANS*; also
  Netherlands Astronomical Satellite), 34 il., 35
Astronomy Missions Board (AMB), NASA,
  53–54
  X-Ray and Gamma-Ray Panel, 54
AT&T (American Telephone & Telegraph Co.),
  75
Atlantic 2 (Intelsat II-C; Intelsat-II F-3), 57
Atlantic 3 (Intelsat-III F-2), 57
Atlantic Missile Range, USAF (AMR; formerly
  Long Range Proving Ground, now Eastern
  Test Range), 149–150
Atlantic Missile Range Operations Office
  (AMROO), MSFC, 149
Atlantic Research Corp., 125
Atlas (intercontinental ballistic missile), 9
Atlas (launch vehicle first stage), 5, 6–7, 8 il., 9
  il., 10, 11–12
Atlas (mythological Greek god), 8 il., 9
Atlas-Able (launch vehicle), 5, 9 il.
Atlas-Agena, 6–7, 9 il.
Atlas-Centaur, 3, 9 il., 11 il.–12, 54
  Atlas-Centaur 2 (development flight), ix
  Atlas-Centaur 3, ix
  Atlas-Centaur 4, ix
Atlas Corp., 9
ATM (Apollo Telescope Mount), 109–110 il.,
  155

Atmosphere Explorer (AE; scientific satellite),
  51 il.
Atmospheric, Magnetospheric, and Plasmas-in-
  Space (AMPS) laboratory, 118
Atmospheric Sounding Projectile (sounding
  rocket). See Asp.
ATS (Applications Technology Satellite), 25,
  36–37, 71
  ATS 1 (ATS-B), ix, 37, 71
  ATS 2 (ATS-A), ix
  ATS 3 (ATS-C), 37 il., 71
  photo by, 30 il.
  ATS 5 (ATS-E), 12
  ATS 6 (ATS-F), 25, 37 il.
Aurora 7 (Mercury-Atlas 7), 108–109
Aurorae (ESRO IA satellite), 45–46 il.
Austria, 117
Auxiliary Flight Research Station, NACA (fore-
  runner of Wallops Flight Center), 158
Azur (GRS-A; German Research Satellite-A;
  later renumbered GRS-A-1), 38, 39 il.

Bailey, Glenn F., 104
Baker, Norman L., 114 n.
Baykonur Cosmodrome, Tyuratam, U.S.S.R.,
  100
Belgium, 117
Bell Aerosystems Co., Textron Corp., 114
Bell Telephone Laboratories, Inc., AT&T, 40, 75
Beltsville Agricultural Research Center, 142
Beltsville Space Center, NASA (renamed God-
  dard Space Flight Center May 1959), 142
Bentsen, Sen. Lloyd M., 147
Beta Centauri (star), 7
Big Joe (launch vehicle), 10 il.
Bios. See Biosatellite.
BIOS (Biological Investigation of Space; reentry
  spacecraft), 38
Biosatellite (biological satellite), 38–39 il., 139
  Biosatellite 1, 39
  Biosatellite 2, 39
  Biosatellite 3, 39
Black Brant (goose), 129
Black Brant (sounding rocket), 123, 128–129
  Black Brant II, 129
  Black Brant IVA, 129, 133
  Black Brant VB, 129
  Black Brant VC, 128 il.–129
Blagonravov, Anatoly A., 103
Blue Scout (launch vehicle), 20
BMwF (German Ministry for Scientific
  Research), 33, 38, 84
Bono, Philip, 113
Boreas (ESRO IB satellite), 46

Bossart, Karel J., 9
Bristol Aerospace Ltd., Canada, 128
British Interplanetary Society, 83, 114
Broglio, Luigi, 67, 188
Brown, John Y., 104
Brucker, Wilbur M., 49
Bumblebee (USN rocket project), 123
Buongiorno, Carlo, 67, 188
Bureau of Ships, USN, 127
Burgess, Eric, 83
Burritt, Elija H., 6–7
Byrd, Sen. Robert C., 147

Cajun (a Louisiana people), 131
Cajun (sounding rocket upper stage), 125, 130
    il., 131, 133
California Institute of Technology (Caltech),
    143–144
Callisto (Jovian moon), 90
Caltech (California Institute of Technology),
    143–144
Canada, 35, 58, 75, 129
Canadian Armament Research and Development
    Establishment (CARDE), 129
Canadian Defence Research Board, 35, 58
Canadian Department of Communication, 58,
    75
"Canary Bird" (nickname for *Intelsat II-C;
    Atlantic 2*), 57 n.
Canary Islands earth station, 57 n.
Cannonball (proposed passive satellite), 58
Cape Canaveral (Cape Kennedy 1963–1973), 16
    il., 17 il., 35, 149
Cape Kennedy. See Cape Canaveral.
CARDE (Canadian Armament Research and
    Development Establishment), 129
Carpenter, M. Scott, 108
CAS (Cooperative Applications Satellite), 41, 58
    CAS-A (*Eole*), 41 il., 53
    CAS-C (renamed CTS-A), 58
Casper (CSM-113, on *Apollo 16*), 100
Castor (mythological Greek twin), 20, 104, 105
    il.
Castor (Scout second stage), 20
Castor (strap-on rocket motor), 13, 15
Centaur (launch vehicle upper stage), ix, 3, 9 il.,
    11 il.–12, 54, 56 il., 71, 152
    name, 11–12
Centaur (mythological man-horse), 11–12 il.
Centaurus (constellation), 6
Centre National d'Études Spatiales (CNES;
    French National Center for Space Studies), 41,
    52, 72

CEPT (Conférence Européene des Postes et
    Télécommunications Satellite; also CETS and
    ECS), 47
Ceres (mythological Greek goddess; proposed
    satellite name), 45
CETS (Conference on European Telecommuni-
    cations Satellite; also CEPT), 47
Chaffee, Roger B., 100
Challenger (LM-12, on *Apollo 17*), 100
Chapman, John, 58
Charles II, King of England, 159
Charlie Brown (CSM-106, on *Apollo 16*), 100
Chesapeake Bay Ecological Test Program, 160
Christian Science Monitor, 113
CIFAS (Consortium Industriel France-Allemand
    pour le Satellite Symphonie), 72
Clark, John F., 37
CM (command module). See Command and
    service module.
CNES (Centre National d'Études Spatiales), 41,
    52, 72
CNR (Consiglio Nazionale delle Ricerche, CNR;
    National Research Council of Italy), 68
Collier's magazine, 113
Columbia (CSM-107, on *Apollo 11*), 98 il., 100
Comision Nacional de Investigación del Espacio
    (CONIE), 54
Command and service module (CSM), x, 98 il.,
    99–100, 104
    CSM-104 (*Gumdrop; Apollo 9*), 100
    CSM-106 (*Charlie Brown; Apollo 10*), 100
    CSM-107 (*Columbia; Apollo 11*), 98 il., 100
    CSM-108 ("*Yankee Clipper*"; *Apollo 12*), x,
        100
    CSM-109 (*Odyssey; Apollo 13*), 100
    CSM-110 (*Kitty Hawk; Apollo 14*), 100
    CSM-111 (ASTP), 101–102 il.–104
    CSM-112 (*Endeavour; Apollo 15*), 100
    CSM-113 (*Casper; Apollo 16*), 100
    CSM-114 (*America; Apollo 17*), 100
Command module (CM). See Command and
    service module.
Commerce, Department of, 77, 158
Committee on Space Research (COSPAR), 167
Communications Satellite Corp., 56
Communications Technology Satellite (CTS), 58
Conférence Européene des Postes et Télécom-
    munications Satellite (CEPT; also CET), 47
Conference on European Telecommunications
    Satellite (CETS; also CEPT), 47
Congress, U.S., 99, 137, 143, 147, 152, 154
    House, 147
        Committee on Public Works, 144 n.

Senate, 147

CONIE (Comision Nacional de Investigación del Espacio, Spain), 55

Consiglio Nazionale delle Ricerche (CNR; National Research Council of Italy), 68

Consolidated Vultee Aircraft Corp. (Convair; later Convair/Astronautics Division, General Dynamics Corp.), 9, 11–12, 22

Consortium Industriel France-Allemand pour le Satellite Symphonie (CIFAS), 72

Constellation (aircraft), 6

Convair (Consolidated Vultee Aircraft Corp.; later Convair/Astronautics Division, General Dynamics Corp.), 9, 11–12, 22

Cooper, L. Gordon, Jr., 107, 109

Cooper Development Corp., 127

Cooperative Applications Satellite (CAS), 41, 53, 58

Copernicus (astronomer), 62

Copernicus (lunar crater), 85 il.

Copernicus (*OAO 3*, Orbiting Astronomical Observatory), 12, 61 il., 62

Cortright, Edgar M., 60, 83, 87, 93, 94, 191

COS-B (Cosmic Ray Observation Satellite), 47

COSPAR (Committee on Space Research), 167

Cross, C. A., 83

CSM. See Command and service module.

CTS-A (Communications Technology Satellite; formerly ISIS-C and CAS-C), 58

Cummings, Clifford D., 90

"Daughter" (ISEE-B International Sun-Earth Explorer), 52

Daytona Beach Operations, NASA, 150

Deacon (rocket motor), 131

Deal, Project, 49

Deep Space Network (DSN), 145

Defense/Space Business Daily (newsletter), 114 n.

Defense, Department of (DOD; see also Advanced Research Projects Agency, Naval Research Laboratory, U.S. Air Force, U.S. Army), 3, 5, 6, 20, 67, 68, 73, 76–77, 112, 115, 131, 167

　Research and Development Board, Committee on Guided Missiles, 9

　Saturn Vehicle Evaluation Committee, represented on, 17 n.

Delta (launch vehicle upper stage), 3, 12–13 il., 15, 22–23 il., 27, 75, 79, 80 il., 143

　name, 22–23

Denmark, 117

DFRC (Dryden Flight Research Center; new name for Flight Research Center), 140 n.

Diana (proposed name for Gemini program), 104

Dieulot, Gerard, 73

Direct Measurement Explorer (DME; scientific satellite), 50, 58

DME (Direct Measurement Explorer; scientific satellite), 50, 58

DOD. See Defense, Department of.

Donnelly, John P., 43 n.

Dornberger, Walter R., 114

Douglas, Donald, Jr., 22

Douglas Aircraft Co., 5, 22, 113

Dryden, Hugh L., 106–107, 140 n.

Dryden Flight Research Center (DFRC; new name for Flight Research Center), 140 n.

DSN (Deep Space Network), 145

Dyna-Soar (hypersonic boost-glide vehicle), 111

Eagle (LM-5, on *Apollo 11*), 98 il., 100

Early Bird (*Intelsat I*), 56–57

Earth, full-disc photo, 30 il.

Earth (proposed satellite name), 45

Earth and ocean physics applications program (EOPAP), 51, 58, 68

Earth Observations Programs, 43

Earth Observatory Satellite (EOS), 45

Earth Resources Experiment Package (EREP), 43

Earth Resources Laboratory, MTF (NSTL after June 1974), 156–158

Earth Resources Observation Program, 43

Earth Resources Observation Satellite (EROS), 43

Earth Resources Program, 43

Earth Resources Survey Program, 43

Earth Resources Survey Satellite (ERS), 43

Earth Resources Technology Satellite (ERTS; renamed Landsat), 15, 42 il.–45

Eastern Test Range, USAF (ETR; formerly Atlantic Missile Range; earlier Long Range Proving Ground), 73, 149–150

Eccentric Geophysical Observatory (EGO; satellite), 63

Echo (passive communications satellite), 15, 40–41

　Echo 1, 21–22, 40 il.–41

　Echo 2, 7, 41

ECS (European Communications Satellite), 47

Edwards Air Force Base (Muroc AFB before Feb. 1950), 140

EGO (Eccentric Geophysical Observatory satellite), 63

Ehricke, Krafft, 11–12

Eisenhower, President Dwight D., 144 n., 154

ELDO (European Launcher Development Organization), 73

Electronics Research Center (ERC), 137, 139 il.–140

Endeavour (CSM-112, on *Apollo 15*), 100

Energy Research and Development Administration (ERDA), 153

ENI (Ente Nazionale Idrocarburi; state-owned oil agency, Italy), 67

Environmental Protection Agency (EPA), 158

Environmental Research Satellite (ERS), 43

Environmental Science Services Administration (ESSA), 48–49, 60, 77

Environmental Survey Satellite (ESSA), 48 il.–49, 60, 77

Eole (CAS-A; FR-2; meteorological satellite), 41 il., 53

EOPAP (earth and ocean physics applications program), 51, 58, 68

EOS (Earth Observatory Satellite), 45
EOS–A, 45
EOS–B, 45

EPA (Environmental Protection Agency), 160

ERC (Electronics Research Center), 137, 139 il.–140

ERDA (Energy Research and Development Administration), 153

EREP (Earth Resources Experiment Package), 43

EROS (Earth Resources Observation Satellite), 43

ERS (Environmental Research Satellite, USAF), 43

ERS (Earth Resources Survey Satellite), 43

ERTS (Earth Resources Technology Satellite; renamed Landsat), 42–45
ERTS 1 (ERTS–A; *Landsat 1*), 15, 42 il.–43
photo by, 44 il.
ERTS–B (Landsat B), 43

Eskimo tribe, 133

ESRO. See European Space Research Organization and ESRO satellites.

ESRO (European Space Research Organization satellite), 45–47
ESRO 1A (*Aurorae*), 45–46 il.
ESRO 1B (*Boreas*), 46
ESRO 2A, 45–46, 57
ESRO 2B (*IRIS 1*), 46–47, 57–58

ESSA (Environmental Science Services Adminis-

tration), 48–49, 60, 77

ESSA (Environmental Survey Satellite), 48–49, 60, 77
ESSA 1, 48, 77
photo from, 48 il.
ESSA 2, 77
ESSA 5, 48 il.

ETR (Eastern Test Range, USAF; formerly Atlantic Missile Range; earlier Long Range Proving Ground), 76, 151–152

Europa II (European launch vehicle), 73

Europe, Western, 67, 72, 97, 114, 116–118, 129

European Communications Satellite (ECS), 47

European Launcher Development Organization (ELDO), 73

European Space Council, 117

European Space Research Organization (ESRO), 45–47, 51, 52, 54, 57–58, 71, 75

EXOSAT (x-ray astronomy satellite), 47

Explorer (balloon), 49

Explorer (scientific satellite), 15, 20–31, 49–52, 63, 66 n., 144, 183
Aeronomy, 49
Air Density, 41, 49–50
Applications, 52
Atmosphere (AE), 51 il.
Direct Measurement (DME), 50, 58
Geodetic (GEOS), 47, 50 il.–52, 64
Heat Capacity Mapping Mission (HCMM), 52
Heavy, 54
High Energy, 54
International Magnetosphere (IME), 52
International Sun-Earth (ISEE), 47, 52
International Ultraviolet (IUE), 47, 52
Interplanetary Monitoring Platform (IMP), 50, 51 il.
Ionosphere (IE), 35, 50
Meteoroid Technology Satellite (MTS), 50
Radio Astronomy (RAE), 50
Small Astronomy Satellite (SAS), 50, 52
Solar, 50
Super, 54
X-ray Astronomy, 50

Explorer 1, 14 il., 15, 49, 50 il., 79, 89, 144, 167

Explorer 2 (ionosphere direct measurement satellite), 35

Explorer 11 (gamma ray astronomy satellite), 50 il.

Explorer 20 (IE-A; "Topsi"; Ionosphere Explorer), 35

Explorer 25 (Injun 4; Air Density Explorer), 52

Explorer 29 (GEOS 1; GEOS-A; Geodetic Satellite), 50 il.–51

INDEX

Explorer 31 (DME-A; Direct Measurement Explorer), 58

Explorer 32 (AE-B; Atmosphere Explorer), 51 il.

Explorer 34 (IMP-F; Interplanetary Monitoring Platform), 51 il.

Explorer 36 (GEOS 2; GEOS-B; Geodetic Satellite), 51

Explorer 40 (Injun 5), 52

Explorer 42 (Uhuru; SAS-A), 50, 68

Explorer 52 (Hawkeye 1), 21, 52

Extended Apollo (Apollo X), 109

FAA (Federal Aviation Administration, 47

Faget, Maxime A., 10, 15–16

Fairchild Stratos Corp., 66

Faith 7 (Mercury-Atlas 9), 109

Falcon (LM-10, on *Apollo 15),* 100

"Fat Albert" (sounding rocket), 127

FCC (Federal Communications Commission), 79

Federal Aviation Administration (FAA), 47

Federal Communications Commission (FCC), 79

Fiorio, Franco, 67, 188

Fletcher, James C., 117, 158

Flight Research Center (FRC; renamed Dryden Flight Research Center, 1976), 137, 140, 141 il.

FOE (Frog Otolith Experiment), 62

FR-1 (FR-1A; French Satellite No. 1), 52–53 il. FR-1B, 53
FR-2. See *Eole.*

France, 41, 52–53, 72–73, 117

FRC (Flight Research Center; renamed Dryden Flight Research Center, DFRC, 1976), 137, 140, 141 il.

Freedom 7 (Mercury-Redstone 3), ix, 16 il., 107–108

French Guiana, 73

"French Satellite." See *Eole.*

"French Satellite No. 1." See *FR-1.*

Friendship 7 (Mercury-Atlas 6), ix, 8 il., 108 il.

Frog Otolith Experiment (FOE), 62

Galactic Jupiter Probe, 90

GALCIT (Guggenheim Aeronautical Laboratory, Caltech; forerunner of Jet Propulsion Laboratory), 143

GALCIT Rocket Research Project, 143–144

GARP (Global Atmospheric Research Project), 71

GATE (GARP Atlantic Tropical Experiment), 71

GATV (Gemini Agena Target Vehicle), 7 il., 25

Gemini (constellation), 20, 104

Gemini (manned space program; Project Gemini), ix–x, 7, 25, 97, 104–106, 147
program names, ix–x, 104–106, 200

Gemini (mythological Greek twins), 104–105 il.–106

Gemini (spacecraft), 25, 104–106, 107 il.
Gemini 3 ("Molly Brown"), 106
Gemini 4, 106
Gemini 6, photo from, 105 il.
Gemini 7, 105 il.
Gemini 8, photo from, 7 il.

Gemini Agena Target Vehicle (GATV), 7 il., 25

Gemini-Titan 1, x, 24 il.

Gemini-Titan 2, x

General Dynamics Corp., Convair/Astronautics Division, 11, 22

General Electric Co., 38

Geodetic Earth Orbiting Satellite. See GEOS (NASA).

Geodetic Explorer Satellite. See GEOS (NASA).

Geodetic Satellite. See GEOS (NASA).

Geodynamic Experimental Ocean Satellite (GEOS-C, NASA), 47, 51–52, 68

Geography of the Heavens (book), 6

GEOS (ESRO Geostationary Scientific Satellite), 47, 52

GEOS (NASA Geodetic Satellite; Geodetic Explorer Satellite; Geodetic Earth Orbiting Satellite), 47, 50–52, 64
GEOS 1 (Explorer 29), 50 il.–51
GEOS 2 (Explorer 36), 51

GEOS-C (NASA Geodynamic Experimental Ocean Satellite), 47, 51–52, 68

Geostationary Operational Environmental Satellite (GOES), 71

Geostationary Scientific Satellite (GEOS, ESRO), 47, 52

German Ministry for Scientific Research (BMwF), 33, 38, 84

German Research Satellite (GRS)
GRS-A (GRS-A-1). See *Azur.*
GRS-A-2. See *Aeros.*

Germany, West (Federal Republic of Germany), 12, 33, 38–39, 72–73, 84, 117, 127

Gesellschaft für Weltraumforschung (GFW; West German Space Agency), 72

GFW. See Gesellschaft für Weltraumforschung.

Gilruth, Robert R., 107

GISS (Goddard Institute for Space Studies), 143

Glenn, John H., Jr., 8, 108 il.

Glennan, T. Keith, 106–107

Global Atmospheric Research Program (GARP), 71

Goddard, Robert H., 2 il., 83, 121, 142 il., 143

Goddard Institute for Space Studies (GISS), 143

Goddard Space Flight Center (GSFC), 35, 73, 84, 124, 137, 142–143 il., 158
  Launch Operations Division, 150
  name, 142–143
  Theoretical Division, 143
  Space Task Group, 104, 107, 147

GOES–A (SMS–C; Geostationary Operational Environmental Satellite), 71

Golden Fleece (Greek mythology), 107, 125

Goldstone antenna, 145 il.

Greb (DOD satellite), 167

Grissom, Virgil I., 100, 105, 108

GRS–A (GRS–A–1). See *Azur*.

GRS–A–2. See *Aeros*.

GSFC. See Goddard Space Flight Center.

Gualtierotti, Torquato, 62

Guggenheim Aeronautical Laboratory, Caltech (GALCIT; forerunner of Jet Propulsion Laboratory), 143

Gumdrop (CSM–104, on *Apollo 9*), 100

Hawaii, 57

Hawk (antiaircraft missile), 131

Hawk (sounding rocket), 130 il., 131, 133

Hawkeye 1 (*Explorer 52*), 21, 52

HCMM (Heat Capacity Mapping Mission; Applications Explorer satellite), 52

HEAO (High Energy Astronomy Observatory), 54
  HEAO–A, 54

Heat Capacity Mapping Mission (HCMM; Applications Explorer satellite), 52

Heavy Explorer. See High Energy Astronomy Observatory.

"Heliocentric" (ISEE–C; International Sun-Earth Explorer), 52

Helios (mythological Greek god), 84

Helios (Advanced Orbiting Solar Observatory, AOSO), 84

Helios (solar probe), 25, 84
  Helios 1, 12, 25, 84 il.
  Helios–B, 84

HEOS (Highly Eccentric Orbit Satellite), 54
  HEOS 1, 47, 54–55 il.
  HEOS 2, 54

Hercules Powder Co., 133

Hermes (mythological Greek god), 106

HEX (Heavy Explorer; High Energy Explorer; proposed names), 54

High Energy A. See High Energy Astronomy Observatory.

High Energy Astronomy Observatory (HEAO), 25, 53 il.–54

High Speed Flight Research Station, NACA (High Speed Flight Station, 1954–1958; HSFS, NASA, Oct. 1958; Flight Research Center, Sept. 1959; Dryden Flight Research Center, Jan. 1976), 140

Highly Eccentric Orbit Satellite (HEOS), 47, 54, 55 il.

Hirsch, Richard, 49

HL–10 (lifting body), 111

Homing All the Way Killer (Hawk antiaircraft missile), 131

Hughes Aircraft Co., 36, 79, 113

Human Factor Systems program, 63

Huntsville Facility (became Marshall Space Flight Center March 1960), 154

Hurricane Alma, 61 il.

IE (Ionosphere Explorer; scientific satellite), 35, 50

IGY (International Geophysical Year), 27, 40, 49, 78–79, 88, 121

IME (International Magnetosphere Explorer) program, 52

IMP (Interplanetary Monitoring Platform; Explorer satellite), 50, 51 il.

Improved Tiros Operational Satellite (ITOS), 49, 60, 71, 76–78

India, 37

Industrial College of the Armed Forces, 104

Injun 1 (scientific satellite), 52
  Injun 2, 52
  Injun 3, 52
  Injun 4 (*Explorer 25*), 52
  Injun 5 (*Explorer 40*), 52
  Injun F (redesignated *Hawkeye 1*; *Explorer 52*), 52

Instituto Nacional de Téchnica Aeroespacial (INTA), 55

INTA (Instituto Nacional de Téchnica Aeroespacial), 55

INTASAT (Spanish ionosphere research satellite), 15, 54–55 il.

INTELSAT (International Telecommunications Satellite Organization; Consortium before Feb. 1973), 56–57

Intelsat (communications satellite), 15, 56–57
  Intelsat I (Early Bird; communications satellite), 56–57
  Intelsat II series, 57
  Intelsat II–A (*Lani Bird*), 57
  Intelsat II–B (*Pacific 1*), 57

INDEX

Intelsat II-C (Intelsat-II F-3; Atlantic 2), 57
Intelsat II-D (Intelsat-II F-4; Pacific 2), 57
Intelsat III series, 57
Intelsat III-A (Intelsat-III F-1), 57
Intelsat-III F-2 (Atlantic 3), 57
Intelsat IV series, 12, 56 il., 57
Intelsat-IV F-2, 57
Intelsat-IV F-8, 57
Interim upper stage (IUS; for Space Shuttle), 115
Interior, Department of the, 42-45, 153, 158
International Council of Scientific Unions, 167
International Geophysical Year (IGY), 27, 40, 49, 78-79, 88, 121
International Magnetosphere Explorer (IME) program, 52
International Radiation Investigation Satellite (*IRIS 1; ESRO 2B*), 46-47, 57-58
International Satellites for Ionospheric Studies (ISIS), 35, 58, 59 il.
International Sun-Earth Explorer (ISEE; scientific satellite), 47, 52
International Telecommunications Satellite Organization (Consortium before Feb. 1973; INTELSAT), 56-57
International Ultraviolet Explorer (IUE; originally SAS-D), 47, 52
International Years of the Quiet Sun, 121
Interplanetary Monitoring Platform (IMP; Explorer satellite), 50, 51 il.
Intrepid (LM-6, on *Apollo 12*), x, 100
Io (Jovian moon), 89 il.
Ionosphere Explorer (IE; scientific satellite), 35, 50
Iowa, Univ. of, 52
Iris (sounding rocket), 58, 125
IRIS 1 (ESRO 2B; International Radiation Investigation Satellite), 46-47, 57-58
ISEE (International Sun-Earth Explorer; scientific satellite), 47, 52
ISEE-A ("Mother"), 52
ISEE-B ("Daughter"), 52
ISEE-C ("Heliocentric"), 52
Isis (ancient Egyptian goddess), 58
ISIS (International Satellites for Ionospheric Studies) project, 35, 58
ISIS 1, 58
ISIS 2, 58, 59 il.
ISIS-C. See CTS-A.
ISIS-X (launch), 58
Italian National Research Council (Consiglio Nazionale delle Ricerche, CNR), 68
Italian Research-Oriented Satellite (SIRIO). 68-69

Italian Space Commission, 67, 188
Italy, 62, 67-68, 117, 188
ITOS (Improved Tiros Operational Satellite), 49, 60, 71, 76-78
ITOS 1 (Tiros-M), 60, 77
ITOS-A (*NOAA 1*), 49, 60, 77
ITOS-B, 76 il.
ITOS-D (*NOAA 2*), 77
ITOS-F (*NOAA 3*), 77
ITOS-G (*NOAA 4*), 15, 55, 77-78
IUE (International Ultraviolet Explorer), 47, 52
IUS (Space Shuttle interim upper stage), 115

Jacobowski, Walter, 94
Jahn, Gunnar, 155 il.
Japan, 71
Jason (Greek mythology), 125
Jason (sounding rocket), 125
Javelin (Argo D-4; sounding rocket), 123, 126 il.-127, 133
Jet Propulsion Laboratory, 15, 49, 89, 90, 121, 137, 143-145
name, 143-145
Johns Hopkins University, 123, 139
Johnson, President Lyndon B., 146 il., 147, 149
Johnson Space Center (JSC; Manned Spacecraft Center before 1973), 103, 137, 146 il., 147, 158
Mission Operations Control Room, 146 il.
name, 147
Jones, Alton E., 73
Jones, Lloyd E., Jr., 67 n.
Journeyman (Argo D-8; sounding rocket), 126 il.-127
JPL. See Jet Propulsion Laboratory.
JSC. See Johnson Space Center.
Juno (launch vehicle), 14-15, 16
Juno I, 14 il., 15, 49
Juno II, 15, 17
Juno III (concept), 17
Juno IV (concept), 17
Juno V, 15, 17
Juno (mythological Roman goddess), 14 il., 15
Jupiter (intercontinental ballistic missile), 15, 16, 17
Jupiter C (Jupiter Composite Reentry Test Vehicle), 15, 17, 49
Jupiter (mythological Roman god), 15
Jupiter (planet), 17, 87, 89 il., 90

Kaesmeier, Karl, 84
Keefer, Eugene C., 12
Keldysh, Mstislav V., 103

Kennedy, President John F., 97, 99, 147, 148 il., 149
Kennedy, Cape. See Cape Canaveral.
Kennedy Space Center (KSC), 105, 137, 148 il., 149–150
  Launch Complex 34, 18 il.
  NASA Daytona Beach Operations, 150
  Vehicle Assembly Building (VAB), 98 il., 148 il.
  Western Test Range Operations Division, 150
Kenya, Africa, 50
Kissinger, Henry A., 103
Kitty Hawk (CSM-110, on *Apollo 14*), 100
Koelle, Heinz H., 111
Kosygin, Premier Aleksey N. (U.S.S.R.), 103
Kronos (proposed name for Saturn V), 20
KSC. See Kennedy Space Center.
Kupperian, James E., Jr., 60

LAGEOS (Laser Geodynamic Satellite), 58–59 il.–60, 68
Landsat (Earth Resources Technology Satellite, new name; see also ERTS)
  Landsat 1 (formerly *ERTS 1*), 43 n.
  Landsat 2 (formerly ERTS-B), 43 n.
Langley, Samuel P., 150, 151 il.
Langley Aeronautical Laboratory, NACA (Langley Memorial Aeronautical Laboratory 1920–1948; Langley Research Center, NASA, Oct. 1958), 20, 40, 140, 150
  Auxiliary Flight Research Station, 158
  Pilotless Aircraft Research Division (PARD), 131, 158
  Station, 159
Langley Research Center (LaRC), 10, 15, 94, 131, 137, 140, 147, 150, 151 il.
  name, 150
Lani Bird (*Intelsat II-A*), 57
LaRC. See Langley Research Center.
Large Space Telescope (LST), 116
Laser Geodynamic Satellite (LAGEOS), 58–59 il.–60
Launch Operations Directorate, MSFC (became Launch Operations Center March 1962; Kennedy Space Center Nov. 1963), 149
  Test Support Office, 152–153
LDEF (Long-Duration Exposure Facility; satellite), 116
LEM (lunar excursion module), 100
Leonov, Aleksey A., 102
LeRC. See Lewis Research Center.
Lewis, George W., 152, 153 il.

Lewis Flight Propulsion Laboratory, NACA (became Lewis Research Center, NASA, Oct. 1958), 152
Lewis Research Center (LeRC), 137, 152 il.–153
  Plum Brook Station, 153
Liberty Bell 7 (*Mercury-Redstone 4*), 105–106, 108
Little Joe (launch vehicle), 10, 14 il., 15–16
  Little Joe II, 16
LM. See Lunar module.
Lockheed Aircraft Corp., 113
  Lockheed Missiles Systems Division (now Lockheed Missiles & Space Co., Inc.), 6–7
Long-Duration Exposure Facility (LDEF), 116
Long Range Proving Ground (became Atlantic Missile Range, then Eastern Test Range), 149
Low, George M., 43 n., 100, 103, 111 n.
LRV (lunar roving vehicle, Rover), 100, 101 il.
LST (Large Space Telescope), 116
LTTAT (long-tank thrust-augmented Thor), 13, 22–23 il.
LTTAT-Delta (long-tank thrust-augmented Thor-Delta), 13, 22–23 il.
Lunar excursion module (LEM), 100
Lunar module (LM), 98 il., 99–100
  LM-3 (*Spider; Apollo 9*), 100
  LM-4 (*Snoopy; Apollo 10*), 100
  LM-5 (*Eagle; Apollo 11*), 98 il., 100
  LM-6 (*Intrepid; Apollo 12*), x, 100
  LM-7 (*Aquarius; Apollo 13*), 100
  LM-8 (*Antares; Apollo 14*), 100
  LM-10 (*Falcon; Apollo 15*), 100
  LM-11 (*Orion; Apollo 16*), 100
  LM-12 (*Challenger; Apollo 17*), 100
Lunar Orbiter (space probe), 83, 84–85 il., 91, 150
  Lunar Orbiter 2, photo by, 85 il.
Lunar roving vehicle (LRV; Rover), 100, 101 il.

M2 (lifting Body), 111
McDonnell Aircraft Corp., 104
McGolrick, Joseph E., 115
MAF (Michoud Assembly Facility), 155–156 il.
Mailgram, 79
Major (missile), 16
Malemute (sounding rocket upper stage), 132 il., 133
Manned Spacecraft Center (MSC). See Johnson Space Center.
Mariner (interplanetary probe), 12, 83, 86–87, 145
  Mariner 2 (P-38), 87

INDEX

Mariner 4 (Mariner D), 87
Mariner 5 (Mariner E), 87
Mariner 6 (Mariner F), 87
Mariner 7 (Mariner G), 87
Mariner 8 (Mariner H), 87
Mariner 9 (Mariner I), 86 il., 87
  photo by, 86 il.
Mariner 10 (Mariner J), 87
  photo by, 86 il.
Mariner Jupiter-Saturn, 87
MAROTS (Maritime Orbital Test Satellite), 47
Mars (planet), x, 15, 25, 87, 93–94
Mars 2 (U.S.S.R. space probe), x
Mars 3, x
Marshall, Gen. George C. (USA), 154–155 il.
Marshall Space Flight Center (MSFC), 66,
  111–112, 116, 137, 149, 154 il.–157
  Launch Operations Directorate, 149, 150
    Atlantic Missile Range Operations Office
      (AMROO), 149
    Missile Firing Laboratory, 149
    Test Support Office, 150
  name, 154–155
"The Martian Probe" (symposium paper), 83
Martin Co., 22, 25
Mathews, Charles W., 43 n.
Mercury (manned space program; Project Mer-
  cury), ix–x, 9, 10, 16, 97, 99, 104, 106–107
  il.–109, 147, 199
  Advanced Mercury (Gemini), 104
  Mercury Mark II (Gemini), 104–105
  program names, ix–x, 106–109, 199
Mercury (mythological Roman god), 106, 108 il.
Mercury (planet), 86 il., 87
Mercury (spacecraft), 10, 14 il., 15, 104–106, 107
  il.–108 il.
  Mercury-Atlas 6 (*Friendship 7*), ix, 8 il., 108 il.
  Mercury-Atlas 7 (*Aurora 7*), 108–109
  Mercury-Atlas 8 (*Sigma 7*), 109
  Mercury-Atlas 9 (*Faith 7*), 109
  Mercury-Redstone 3 (*Freedom 7*), ix, 16 il.,
    107–108
  Mercury-Redstone 4 (*Liberty Bell 7*), 105–106,
    108
Merritt Island Launch Area (MILA), NASA, 149
Meteoroid Technology Satellite (MTS; Ex-
  plorer), 50
METEOSAT (meteorological satellite), 47
Michoud Assembly Facility (MAF), 155–156 il.
Michoud Operations (renamed Michoud Assem-
  bly Facility July 1965), 155
Michoud Ordnance Plant, 155

MILA (Merritt Island Launch Area, NASA), 149
Milan, University of, 62
Minuteman (intercontinental ballistic missile),
  20, 127
Miranda (UK–X4; experimental satellite), 36
Mississippi Test Facility (MTF; became National
  Space Technology Laboratories June 1974),
  156–157 il.–158
Mississippi Test Operations (renamed Mississippi
  Test Facility July 1965), 156
Moffett Field Laboratory, NACA (forerunner of
  Ames Research Center), 137
"Molly Brown" (*Gemini 3*), 105–106
"Mother" (ISEE–A; International Sun-Earth
  Explorer), 52
MSC (Manned Spacecraft Center). See Johnson
  Space Center.
MSFC. See Marshall Space Flight Center.
MTF (Mississippi Test Facility), 159–161
MTS (Meteoroid Technology Satellite; Ex-
  plorer), 50
Mueller, George E., 19, 114
Muroc Air Force Base (Edwards AFB after Feb.
  1950), 140
Muroc Flight Test Unit, NACA (forerunner of
  Flight Research Center), 140
Muroc test facility, USAF, 140
MX-1593 (USAF Atlas project), 9

NACA. See National Advisory Committee for
  Aeronautics.
Nagy, Alex P., 104, 200
NAS (National Academy of Sciences), 62
NASA/Industry Program Plans Conference, 99
NASC (National Aeronautics and Space Coun-
  cil), 147
National Academy of Sciences (NAS), 62, 167
National Advisory Committee for Aeronautics
  (NACA), 20, 40, 49, 111, 132, 139, 140, 150,
  152, 158–159
National Aeronautics and Space Council
  (NASC), 147
National Defense Research Council, 131
National Geographic Society, 49
National Launch Vehicle Program, 3
National Oceanic and Atmospheric Administra-
  tion (NOAA), 48–49, 60, 68, 71, 77, 153
National Operational Environmental Satellite
  System, 71 n.
National Operational Meteorological Satellite
  System (NOMSS), 71
  National Science Foundation, 153

National Security Council, Ad Hoc Committee on Outer Space, 49
National Space Club, 114
National Space Technology Laboratories (NSTL; Mississippi Test Facility before June 1974), 137, 157 il.–158
Natural Resources Program, 43
Naugle, John E., 87 n.
Naval Research Laboratory (NRL), 79, 127, 142
Netherlands, 35, 117
    Ministry of Economic Affairs, 35
    Ministry of Education and Science, 35
Netherlands Astronomical Satellite (Astronomical Netherlands Satellite, ANS), 34 il., 35
Newell, Homer E., 36–37
Nike (antiaircraft missile), 133
Nike (mythological Greek goddess), 132 il., 133
Nike (sounding rocket first stage), 124 il., 127, 128 il., 130 il., 131, 132 il., 133
Nike-Apache (sounding rocket), 123, 124 il., 133
Nike-Asp, 127, 128 il.
Nike-Cajun, 123, 124, 130 il., 133
Nike-Hawk, 123, 131, 133
Nike-Javelin, 127
Nike-Malemute, 123, 132 il., 133
Nike-Tomahawk, 123, 132 il., 133
Nimbus (meteorological satellite), 7, 60
    Nimbus 1 (Nimbus A), 60, 61 il.
    Nimbus 2 (Nimbus C), 60
    Nimbus 3 (Nimbus B–2), 60
    Nimbus 4 (Nimbus D), 60
    Nimbus 5 (Nimbus E), 60
Nix Olympica (Martian mountain), 92 il.
Nixon, President Richard M., 103, 114–115, 116, 147
NOAA. See National Oceanic and Atmospheric Administration and NOAA satellites.
NOAA (National Oceanic and Atmospheric Administration satellite), 49, 60, 77
    NOAA 1 (ITOS-A), 49, 60, 77
    NOAA 2 (ITOS-D), 77
    NOAA 3 (ITOS-F), 77
    NOAA 4 (ITOS-G), 15, 55, 77–78
Nobel Peace Prize, 155
NOMSS (National Operational Meteorological Satellite System), 71
North Eastern Office, NASA, 140
NRL (Naval Research Laboratory), 79, 127, 142
NSTL (National Space Technology Laboratories), 137, 157 il.–158

OA (Office of Applications, NASA), 43 n., 58

OAO (Orbiting Astronomical Observatory), 60–62
    OAO 1 (OAO-A1), 62
    OAO 2 (OAO-A2), ix, 12, 62
    OAO 3 (*Copernicus;* OAO-C), 12, 61 il., 62
OART (Office of Advanced Research and Technology, NASA), 37
Ocean (a Titan, Roman mythology), 25 il.
Odyssey (CSM-109, on *Apollo 13*), 100
Office of Advanced Research and Technology (OART), NASA, 37
Office of Applications (OA), NASA, 43 n., 58
Office of International Affairs, NASA, 58
Office of Launch Vehicles, NASA, 115
Office of Manned Space Flight (OMSF), NASA, 19, 100, 104–105
Office of Space Science and Applications (OSSA), NASA, 36–37
Office of Space Vehicle Research and Technology, NASA, 66
OFO (Orbiting Frog Otolith satellite), 62 il.–63
OGO (Orbiting Geophysical Observatory), 7, 63 il.
Ohio National Guard, 153
Olympia (proposed West German satellite), 73
OMSF (Office of Manned Space Flight, NASA), 19, 100, 104–105
OOS (Space Shuttle orbit-to-orbit stage), 115
Orbit-to-orbit stage (OOS; for Space Shuttle), 115
"Orbital Radio Relays" (article), 40
Orbital Test Satellite (OTS), 47
Orbital Workshop. See Skylab.
Orbiter. See Lunar Orbiter, Pioneer, Space Shuttle, Viking (Mars probe).
Orbiting Astronomical Observatory (OAO), ix, 12, 60–61 il.–62
Orbiting Frog Otolith (*OFO*) satellite, 62 il.–63
Orbiting Geophysical Observatory (OGO), 7, 63 il.
Orbiting Observatory series, 60–64
Orbiting Satellite Carrying Amateur Radio (OSCAR), 15, 55
Orbiting Solar Observatory (OSO), 15, 63–64 il.
Orion (LM-11, on *Apollo 16*), 100
Orpheus (proposed name for Gemini program), 104
OSCAR (Orbiting Satellite Carrying Amateur Radio)
    OSCAR 7 (*AMSAT OSCAR 7*), 15, 55
OSO (Orbiting Solar Observatory), 15, 63–64
    OSO 1 (S–16), 64

OSO 3 (OSO–E), 64 il.

OSO 7 (OSO–H), 64

OSSA (Office of Space Science and Applications, NASA), 36–37

O'Sullivan, William J., Jr., 40–41

OTS (Orbital Test Satellite), 47

Ousley, Gilbert W., 84

Pacific 1 (*Intelsat II–B*), 57

Pacific 2 (*Intelsat II–D; Intelsat-II F–4*), 57

Pacific Launch Operations Office (PLOO), NASA, 150

Pacific Missile Range, USAF (PMR; now Western Test Range), 35, 150

Test Support Office, Launch Operations Directorate, 150

PAGEOS (Passive Geodetic Earth Orbiting Satellite) project, 64–65

Pageos 1, 41, 65 il.

Paine, Thomas O., 103, 116–117

PARD. See Pilotless Aircraft Research Division.

Passive Geodetic Earth Orbiting Satellite (PAGEOS), 41, 64–65 il.

Pearl River Test Site (early designation for Mississippi Test Facility), 156 n.

Pegasus (meteoroid satellite), 65 il.–67

Pegasus 1, 67

Pegasus 2, 67

Pegasus 3, 67

Pegasus (winged horse, Greek mythology), 66

Perseus (constellation), 20

Pickering, William H., 15, 90

Pierce, John R., 40

Pilotless Aircraft Research Division (PARD), Langley Aeronautical Laboratory, 131, 158–159

Pilotless Aircraft Research Station (became Wallops Station, NASA, Oct. 1958), 159

Pioneer (space probe), 12, 15, 83, 88–90, 139

Pioneer 1, ix, 88–89

Pioneer 2, 89

Pioneer 3, ix, 89, 144

Pioneer 4, 89, 144

Pioneer 5 (F–3), 89

Pioneer 6 (Pioneer A), 89

Pioneer 7 (Pioneer B), 89–90

Pioneer 8 (Pioneer C), 89–90

Pioneer 9 (Pioneer D), 89–90

Pioneer 10 (Pioneer F), 90

photo by, 89 il.

Pioneer 11 (Pioneer G), 88 il., 90

Pioneer Saturn. See *Pioneer 11.*

Pioneer Venus, 90

Planetary Programs Office, NASA, 94

PLOO (Pacific Launch Operations Office, NASA), 150

Plum Brook Station, LeRC, 153

Space Power Facility, 153

PMR (Pacific Missile Range, USAF; now Western Test Range), 35, 150

POGO (Polar Orbiting Geophysical Observatory), 63

Polar Orbiting Geophysical Observatory (POGO), 63

Polaris (intercontinental ballistic missile), 7, 20

Pollux (mythological Greek twin), 104, 105 il.

Pollux (rocket motor), 16

Pressly, Eleanor, 125

Project Designation Committee, NASA, 183–184

launch vehicle names, 19, 20

manned space flight names, 100, 103, 109, 111 n., 116

satellite names, 38, 43 n., 44–45, 52, 66

space probe names, 94, 183–184

Purser, Paul E., 16

Radio Astronomy Explorer (RAE; scientific satellite), 50

Radio Corporation of America (now RCA Corp.), 76

RAE (Radio Astronomy Explorer; scientific satellite), 50

RAM (Research and Applications Module; Spacelab forerunner), 116

Ranger (lunar probe), 83, 84, 90–91, 93, 145

Ranger 4 (P–35), 91

Ranger 7 (Ranger B), 91 il.

Ranger 8 (Ranger C), 91

Ranger 9 (Ranger D), 91

photo by, 91 il.

Ranger (program; Project Ranger), 84, 90–91, 145

Ranger (truck), 90

RCA Corp., 76

"The Recoverable, Reusable Space Shuttle" (address), 114

Recruit (rocket), 16, 22

Redstone (intermediate-range ballistic missile), 16, 149

Redstone (launch vehicle first stage), ix, 15, 16 il., 105–108

Redstone Arsenal, 16, 154

Relay (active repeater communications satellite),
15, 31, 66 il.
Relay 1 (A–15), 67
Relay 2 (A–16), 67
Rendezvous (periodical), 114
"Rendock" (rendezvous and docking mission),
103
Research and Applications Module (RAM;
Spacelab forerunner), 116
Reusable Orbital Module Booster and Utility
Shuttle (ROMBUS), 113
Rockwell International Corp., 111 n.
Rogers, William P., 103
Roman, Nancy G., 60
ROMBUS (Reusable Orbital Module Booster
and Utility Shuttle), 113
Roosevelt, President Franklin D., 139
Rosen, Josephine (Mrs. Milton W.), 79
Rosen, Milton W., 12, 79
Rover (lunar roving vehicle; LRV), 100, 101 il.
Rowland, Joe, 22, 25

S–I (Saturn I first stage), 19, 187
S–IB (Saturn IB first stage), 19, 155
S–IC (Saturn V first stage), 20, 155–156 il.–158
S–II (Saturn V second stage), 20, 156–158
S–IV (Saturn I second stage), 19, 187
S–IVB
   Saturn IB second stage, 19, 20, 109
   Saturn V third stage, 20, 109
S–V (Saturn C–1 stage), 187
Saint Mark, 67
Saliga, Stephen A., 88
San Marco launch platform, 50, 67–68
San Marco (U.S.-Italy satellite project), 20, 22,
67–68, 188
   San Marco 1 (Italian satellite), 66 il., 67–68
   San Marco 2, 67–68
   San Marco 3, 68
   San Marco 4, 68
Sandia Laboratories, 127, 133
SAROS (Satellite de Radiodiffusion pour Orbit
Stationnaire), 73
SAS (Small Astronomy Satellite; Explorer), 50,
52
   SAS-A (*Explorer 42; Uhuru*), 50, 68
   SAS-D (International Ultraviolet Explorer;
IUE), 47, 52
Satellite de Radiodiffusion pour Orbit Station-
naire (SAROS), 73
Satellite Italiano Ricerche Orientate (SIRIO),
68–69

Saturn (launch vehicle), 17–21, 58, 109, 155–158
   Advanced Saturn (see also Saturn V), 20
   Saturn A, 19
   Saturn B, 19
   Saturn C, 19
      C–1 (see also Saturn I), 19, 187
      C–1B (see also Saturn IB), 19, 187
      C–3, 19
      C–4, 19
      C–5 (see also Saturn V), 19–20
   Saturn I, x, 15, 17 il., 19, 66, 109
      Uprated Saturn I (see also Saturn IB), 19
   Saturn IB, x, 3, 17–18 il., 19, 20, 100, 104, 155
   Saturn V, x, 3, 18 il., 19–20, 94 n., 100, 109,
111–112, 148 il., 155–158
Saturn (mythological Roman god), 17, 18 il.
Saturn (planet), 17, 87, 90
Saturn Apollo, x
Saturn Vehicle Evaluation Committee (Silver-
stein Committee), 17–19
Schilling, G. F., 60
Schirra, Walter M., Jr., 109
Schumann, Maurice, 73
Schumann, Robert, 73
Science Advisory Committee, President's, 113
Scorpio (constellation), 21
Scout (launch vehicle), 3, 20–21 il., 35, 38, 47,
150, 160
   Blue Scout, 20
   San Marco Scout 1, 67
   Scout E, 21
Seamans, Robert C., Jr., 20, 37, 104, 200
SEASAT (Specialized Experimental Applica-
tions Satellite), 43 n., 68–69 il.
SEASAT User Working Group, 68
SEOS (Synchronous Earth Observatory
Satellite), 45
Sergeant (launch vehicle first stage), 22
Service module (SM). See Command and service
module.
SEX (Super Explorer, proposed name), 54
Sheldon, Courtney, 113
Shepard, Alan B., Jr., 16, 108
Shotput (launch vehicle), 21 il.–22, 40, 188
Sigma 7 (*Mercury-Atlas 8*), 109
Silverstein, Abe, 17 n., 67, 99, 106
Silverstein Committee (Saturn Vehicle Evalua-
tion Committee), 17–19
Simpson, George L., Jr., 20
SIRIO (Satellite Italiano Ricerche Orientate),
68–69
Skylab (Orbital Workshop program; Project
Skylab), 19, 43, 68, 97, 109–111, 129, 147,

155, 158
name, 109–111
Skylab 1 (SL–1; Orbital Workshop), 20, 97, 109–110 il.–111, 129, 155
Skylab 2 (SL–2; manned mission), 19, 109
Skylab 3 (SL–3), 19, 111
Skylab 4 (SL–4), 19, 111
Skynet (U.K. military communications satellite), 36
  Skynet I series, 36
  Skynet II series, 36
SL–1, SL–2, etc. See *Skylab 1,* etc.
Slayton, Donald K., 102 il.
SM (service module). See Command and service module.
Small Astronomy Satellite (SAS; Explorer), 50, 52
Smith, Rep. H. Allen, 144 n.
Smith, H. Allen, Jet Propulsion Laboratory (proposed name), 144 n.
Smithsonian Institution, 150
SMS (Synchronous Meteorolgical Satellite), 33, 69–71
  SMS 1 (SMS–A), 71
  SMS–B, 70 il., 71
  SMS–C (GOES–A), 71
Snoopy (LM–4, on *Apollo 10*), 100
Solar Explorer (scientific satellite), 50
Sortie Can (Spacelab forerunner), 116
Sortie Lab. See Spacelab.
Soviet Academy of Sciences, 103
Soyuz (Soviet spacecraft), 101–102 il.–104
Space Flight Research Center (proposed), 159
Space Plasma High Voltage Interaction Experiment satellite (SPHINX), 71
Space Shuttle, 3, 45, 54, 58, 97, 111–112 il.–113 il., 118, 147, 150, 155–158
  Conference, 114, 117
  name, 111–114
  Steering Group, 114
  Task Group, 114
"Space Shuttle of the Future: The Aerospaceplane" (article), 114
Space Task Group, NASA (forerunner of Manned Space Center), 104, 107, 147
Space Task Group, President's, 101, 114, 115, 116
Space Technology Laboratories, Inc., 5, 89
Space Tracking and Data Acquisition Network (STADAN; now STDN), 143
Space Transporter (European proposal), 114
Space Tug, 115–116
Spaceflight Tracking and Data Network (STDN; formerly STADAN), 143
Spacelab, 97, 115, 116–117 il.–118

Spaceplane (proposed name), 111 n., 115 n.
Spain, 54–55, 117
Specialized Experimental Applications Satellite (SEASAT), 43 n., 68–69 il.
SPHINX (Space Plasma High Voltage Interaction Experiment satellite), 71
Spider (LM–3, on *Apollo 9*), 100
Sputnik I (U.S.S.R. satellite), 49, 167
STADAN (Space Tracking and Data Acquisition Network; now STDN), 143
START (USAF reentry program), 114
STDN (Spaceflight Tracking and Data Network; formerly STADAN), 143
Steelman, Donald L., 109
Stoney, William E., Jr., 20
Super Explorer. See High Energy Astronomy Observatory.
Survey (proposed satellite name), 45
Surveyor (lunar probe), 12, 83, 84–85, 91, 92 il., 93, 145
  Surveyor 1 (Surveyor A), 93
  Surveyor 3 (Surveyor C), 93
  Surveyor 5 (Surveyor E), 92 il., 93
  Surveyor 6 (Surveyor F), 93
  Surveyor 7 (Surveyor G), 93
Surveyor Orbiter, 85
Switzerland, 117
Symphonie (Franco-German communications satellite), 72 il.–73
  Symphonie 1 (Symphonie–A), 72 il., 73
  Symphonie B, 73
Synchronous Earth Observatory Satellite (SEOS), 45
Synchronous Meteorological Satellite (SMS), 33, 69–70 il.–71
Syncom (synchronous communications satellite), 15, 72 il., 73
  Syncom 1 (A–25), 73
  Syncom 2 (A–26), 73
  Syncom 3 (A–27), 73

TAD (thrust-augmented Delta). See Thor-Delta.
TAID (thrust-augmented improved Delta). See Thor-Delta.
TAT (thrust-augmented Thor). See Thor and Thor-Delta.
TAT-Delta. See Thor and Thor-Delta.
TD (ESRO Thor-Delta solar astronomy satellite) project, 75
  TD–1 (satellite), 75
  TD–2, 75
  TD–1A, 47, 74 il., 75
TE–364–4 (Thor-Delta third stage), 13

Telesat (Canadian telecommunications satellite), 75

Telesat-A (*Anik 1*), 74 il., 75

Telesat-B (*Anik 2*), 75

Telesat-C (*Anik 3*), 75

Telesat Canada (corporation), 75

Television and Infra-Red Observation Satellite (TIROS, later Tiros; meteorological satellite), vii, 7, 15, 31, 48, 69, 76–78

name, vii, 76–78

Telstar (active repeater communications satellite), 15, 75

Telstar 1 (A–40), 23 il., 75 il.

Telstar 2 (A–41), 75

The Tempest (play), 35

Tennessee, University of, Space Institute, 114

Terrier (antiaircraft missile), 133

Terrier (sounding rocket first stage), 133

Terrier-Malemute (sounding rocket), 123, 133

Textron Corp., Bell Aerosystems Co., 114

Thibodaux, Joseph G., Jr., 131

Thiokol Chemical Corp. (later Thiokol Corp.), 125, 133

Elkton Division, 131

Thor (intermediate-range ballistic missile), 22

Thor (launch vehicle first stage), 5 il., 6 il., 7, 12–13 il., 15, 22–23 il.

name, 22–23

thrust-augmented (TAT), 12, 13 il., 22

long-tank thrust-augmented (LTTAT, or Thorad), 13, 22–23 il.

Thor (mythological Norse god), 22, 23 il.

Thor-Able (launch vehicle), 5 il., 12, 188

Thor-Agena, 6 il.–7, 22, 58

thrust-augmented, 22

Thor-Delta, 3, 12–13 il., 15, 22–23 il., 75, 79, 143

name, 12–13, 22–23

long-tank thrust-augmented (LTTAT–Delta), 13, 22–23 il.

Straight Eight, 13 il.

Super Six, 13

thrust-augmented (TAD), 12, 22

thrust-augmented improved (TAID), 13 il.

Thorad (long-tank thrust-augmented Thor). See Thor.

Thorad-Agena, 22

Tinnan, Leonard M., 112

Tiros (originally TIROS; Television and Infra-Red Observation Satellite), vii, 7, 15, 31, 48, 60, 69, 76–78

name, vii, 76–77

Tiros 1 (A–1) through *Tiros 8* (A–53), 77

Tiros 9 (Tiros-I [eye]), 77

Tiros 10 (OT–1), 77

Tiros-K, 71

Tiros-M (*ITOS 1*), 60

Tiros Operational Satellite (TOS; meteorological satellite), 15, 48, 49, 76–77

Titan (giant, Roman mythology), 25 il.

Titan (intercontinental ballistic missile), 22, 25

Titan I, 25

Titan II, 25

Titan (launch vehicle), 24–25

Titan II, 24 il.

Titan III, 12, 25

IIIC, 25

IIID, 25

IIIE, 12, 24 il., 25, 54

Titan-Centaur (launch vehicle), 3, 25, 71

Titan IIIE-Centaur, 12, 24 il., 84, 94

Tomahawk (sounding rocket upper stage), 132 il., 133

Top Kick (proposed satellite name), 49

"Topsi" (*Explorer 20;* IE-A Ionosphere Explorer), 25

Topside sounder (ionospheric satellite), 35, 58

TOS (Tiros Operational Satellite), 15, 48–49, 76–77

Improved TOS. See ITOS.

Transit II-A (DOD satellite), 167

Transportation, Department of, 140, 158

Transportation Development Center (formerly Electronics Research Center), 140

UE (Redstone No. 29), 15

Uhuru (*Explorer 42;* Small Astronomy Satellite A), 50, 68

UK-1 though UK-5. See Ariel.

UK-X4 (*Miranda*), 36

United Kingdom, 35–36, 47, 52

Ministry of Defence, 36

United Nations, 167

The Unsinkable Molly Brown (musical comedy), 105

Uranus (planet), 90

Ursa (missile), 16

U.S. Air Force, 153

Air Research and Development Command (ARDC), 22, 25

Ballistic Missile Division (AFBMD), 88–89

Cambridge Research Laboratories (AFCRL), 133

Edwards Air Force Base, 140

launch vehicles and missiles, 6, 9, 12, 17 n., 20, 22, 25

INDEX

Muroc Air Force Base (Edwards AFB after Feb. 1950), 140

satellites and probes, 35–36, 43, 52, 88–89

sounding rockets, 123, 129, 133

Space Shuttle, 111–114, 115

U.S. Army (see also Army Ballistic Missile Agency, ABMA), 15, 16, 49, 89, 131, 153, 158

   Ordnance, 121, 133, 144, 155

      Guided Missile Center, 154

         Experimental Missiles Firing Branch, 149

         Missile Firing Laboratory (transferred to MSFC, 1960), 149

         Plum Brook Ordnance Works, 153

U.S. Army Air Forces, 159

U.S. Army Air Service, 49

U.S. Navy (see also Naval Research Laboratory), 123, 129, 133, 153

   Bureau of Aeronautics, 159

   Bureau of Ships, 127

U.S. Weather Bureau (USWB), 77

USNS *Croatan*, 128 il.

U.S.S.R. (Union of Soviet Socialist Republics), x, 19, 49, 71, 97, 101–104

USWB (U.S. Weather Bureau), 77

V–2 (German rocket), 149

VAB (Vehicle Assembly Building, KSC), 98 il., 148 il.

Valiant (proposed name for Gemini program), 104

Van Allen, James A., 123

Vanguard (launch vehicle), 5, 12, 26 il.–27, 78, 79, 188

Vanguard (program; Project Vanguard), 27, 78–79, 142

   Vanguard 1 (scientific satellite), 79, 167

   Vanguard 2, 78 il.–79

   Vanguard 3, 79

Vehicle Assembly Building (VAB), KSC, 98 il., 148 il.

Venus (planet), 15, 25, 87, 90

Viking (launch vehicle first stage), 79, 93 n., 150

Viking (Mars probe), 25, 71, 83, 93 il.–94

Viking (Navy sounding rocket), 93 n.

von Braun, Wernher, 17, 113, 154

von Kármán, Theodore, 144

Voyager, Project, 93, 94 n.

Wac Corporal (sounding rocket), 121

Wallop, John, 159

Wallops Flight Center (WFC; Wallops Station before April 1974), 21 il., 67, 131, 137, 158–159 il.–160

   high-gain antenna, 120 il.

   name, 158–160

   Wallops Main Base, 159

Wallops Station (WS). See Wallops Flight Center.

Webb, James E., 19

Westar (Western Union communications satellite), 79–80

   Westar 1 (Westar–A), 79–80 il.

   Westar 2 (Westar–B), 80

   Westar–C, 80

Western Test Range (WTR; formerly Pacific Missile Range, USAF), 58, 150

Western Union Telegraph Co., 79–80

White, Edward H., II, 100, 106 il.

White Sands Missile Range, USA, 129, 147–149

White Sands Test Facility, NASA (White Sands Operations before June 1965), 147–149

Williams, Allen E., 131

Williams College, 68

Wilson, President Woodrow, 150

Wright-Patterson Air Force Base, 88

WS (Wallops Station). See Wallops Flight Center.

WTR (Western Test Range; formerly Pacific Missile Range, USAF), 58, 150

X–1 (rocket research aircraft), 140

X–15 (rocket research aircraft), 111, 140, 141 il.

X–24 (lifting body), 111, 140

X–248 (launch vehicle upper stage), 22, 199 n.

X-ray Astronomy Explorer (scientific satellite), 50

Yankee Clipper (CSM-108, on *Apollo 12*), x, 100

Young, John W., 105

THE AUTHORS

Helen T. Wells, editor in the NASA History Office from 1962 to 1967, came to NASA Headquarters from two years in the Historical Office of Marshall Space Flight Center, an office which she helped establish. She has also served as editorial information specialist in the Historical Office of the Army Ballistic Missile Agency, Redstone Arsenal, Alabama. A graduate of Mary Washington College of the University of Virginia, Mrs. Wells took her B.A. degree in English literature.

Susan H. Whiteley, writer in the NASA History Office 1971–1972, previously was an editorial assistant with the Association for Childhood Education International and an associate editor of the American University literary magazine, the *American*. The *American* was voted the best U.S. collegiate magazine in 1968 by the national professional journalism fraternity.

Carrie E. Karegeannes, editor in the NASA History Office since 1968, has been research analyst, writer, and editor for the nonprofit research corporation Analytic Services Inc., the Central Intelligence Agency, and the U.S. Economic Cooperation Administration's European Headquarters. Earlier she was a reporter for United Press Associations and newspaper reporter and editor in South Carolina and Paris, following receipt of her B.A. degree in English and journalism from Baylor University.

NASA HISTORICAL PUBLICATIONS

HISTORIES

Frank W. Anderson, Jr., *Orders of Magnitude: A History of NACA and NASA, 1915-1976*, NASA SP-4403, 1976, GPO.*

William R. Corliss, *NASA Sounding Rockets, 1958-1968: A Historical Summary*, NASA SP-4401, 1971, GPO.

Constance McL. Green and Milton Lomask, *Vanguard—A History*, NASA SP-4202, 1970; also Washington: Smithsonian Institution Press, 1971.

Barton C. Hacker and James M. Grimwood, *On the Shoulders of Titans: A History of Project Gemini*, NASA SP-4203, in press.

Edwin P. Hartman, *Adventures in Research: A History of the Ames Research Center, 1940-1965*, NASA SP-4302, 1970, NTIS.**

Mae Mills Link, *Space Medicine in Project Mercury*, NASA SP-4003, 1965, NTIS.

Alfred Rosenthal, *Venture into Space: Early Years of Goddard Space Flight Center*, NASA SP-4301, 1968, NTIS.

Robert L. Rosholt, *An Administrative History of NASA, 1958-1963*, NASA SP-4101, 1966, NTIS.

Loyd S. Swenson, James M. Grimwood, and Charles C. Alexander, *This New Ocean: A History of Project Mercury*, NASA SP-4201, 1966, NTIS.

REFERENCE WORKS

The Apollo Spacecraft: A Chronology, NASA SP-4009: Volume I, 1969, NTIS. Volume II, 1973, GPO. Volume III, in press. Volume IV, forthcoming.

Astronautics and Aeronautics: A Chronology of Science, Technology, and Policy, annual volumes from 1961, with an earlier summary volume, *Aeronautics and Astronautics, 1915-1960*. Early volumes available from NTIS; recent volumes from GPO. *Astronautics and Aeronautics, 1973*, NASA SP-4018, 1975.

Katherine M. Dickson (Library of Congress), *History of Aeronautics and Astronautics: A Preliminary Bibliography*, NASA HHR-29, NTIS.

Project Gemini Technology and Operations: A Chronology, NASA SP-4002, 1969, NTIS.

Project Mercury: A Chronology, NASA SP-4001, 1963, NTIS.

Project Ranger: A Chronology, JPL/HR-2, 1971, NTIS.

Skylab: Preliminary Chronology, NASA HHN-130, May 1973, NTIS.

Jane Van Nimmen and Leonard C. Bruno, *NASA Historical Data Book, 1958-1968*, Vol. I, *NASA Resources*, NASA SP-4012, 1976, NTIS.

---

*GPO: Available from Superintendent of Documents, Government Printing Office, Washington, D.C. 20402.

**NTIS: Available from National Technical Information Service, Springfield, Virginia 22161.